Dynamical Systems and Nonlinear Waves in Plasmas

Asit Saha
Department of Mathematics
Sikkim Manipal Institute of Technology
Sikkim Manipal University
East-Sikkim, India

Santo Banerjee
Department of Mathematical Sciences
Politecnico di Torino, Turin, Italy

CRC Press
Taylor & Francis Group
Boca Raton   London   New York

CRC Press is an imprint of the
Taylor & Francis Group, an **informa** business
A SCIENCE PUBLISHERS BOOK

First edition published 2021
by CRC Press
6000 Broken Sound Parkway NW, Suite 300, Boca Raton, FL 33487-2742

and by CRC Press
4 Park Square, Milton Park, Abingdon, Oxon OX14 4RN

© 2021 Taylor & Francis Group, LLC
CRC Press is an imprint of Taylor & Francis Group, an Informa business

Reasonable efforts have been made to publish reliable data and information, but the author and publisher cannot assume responsibility for the validity of all materials or the consequences of their use. The authors and publishers have attempted to trace the copyright holders of all material reproduced in this publication and apologize to copyright holders if permission to publish in this form has not been obtained. If any copyright material has not been acknowledged please write and let us know so we may rectify in any future reprint.

Except as permitted under U.S. Copyright Law, no part of this book may be reprinted, reproduced, transmitted, or utilized in any form by any electronic, mechanical, or other means, now known or hereafter invented, including photocopying, microfilming, and recording, or in any information storage or retrieval system, without written permission from the publishers.

For permission to photocopy or use material electronically from this work, access www.copyright.com or contact the Copyright Clearance Center, Inc. (CCC), 222 Rosewood Drive, Danvers, MA 01923, 978-750-8400. For works that are not available on CCC please contact mpkbookspermissions@tandf.co.uk

Trademark notice: Product or corporate names may be trademarks or registered trademarks and are used only for identification and explanation without intent to infringe.

ISBN: 978-0-367-48732-4 (hbk)
ISBN: 978-1-032-02568-1 (pbk)
ISBN: 978-1-003-04254-9 (ebk)

Typeset in Times New Roman
by Radiant Productions

Preface

This book is aimed at postgraduate students and researchers in Physics and Mathematics, especially in the areas of dynamical systems, nonlinear waves and waves in plasmas. The goal is to explain the dynamics of nonlinear waves and their properties in plasmas, and to show how a dynamical system can be used to understand exciting features, such as solitary wave, periodic wave, supernonlinear wave, chaotic, quasiperiodic and coexisting structures of nonlinear waves in plasmas. The multistability is a fascinating phenomenon of a dynamical system which has been recently explored. Many chaotic systems have several possible final stable states (attractors) for a given set of parameters. This nonlinear phenomenon is known as multistability behavior or coexisting attractors. More interestingly, multistability behaviors have been observed in numerous natural systems and usually play a crucial role in their performance. The clear evidence of multistability behavior was first manifested experimentally in a Q-switched gas laser, since then, chaotic models with multistability behaviors have been extensively reported in both continuous and discrete systems. Of late, the phenomenon have been investigated in various plasma models. Regarding the nonlinear discharge theory, multistability in plasmas can be an interesting basis of experimental observations. Researchers can work experimentally on the multistable phenomena in plasmas, including the condition for coexisting attractors, the existence of chaos and its various related applications.

The theory of planar dynamical system is developed systemically starting with first-order differential equations and their bifurcations, phase plane analysis, and time series plots under the framework of different nonlinear evolution equations. The nonlinear evolution equations which will be covered in this book are the KdV equation, Burgers equation, KdV-Burgers equation, KP equation and Schrodinger equation. Analytical forms of the wave solutions, such as solitary and periodic wave solutions, shock wave solution, superperiodic wave solution are presented using the theory of planar dynamical systems.

The unique feature of the book is to emphasize on different kinds of nonlinear and supernonlinear waves with small and arbitrary amplitudes in plasmas. These include ion-acoustic wave, dust-acoustic wave, dust-ion acoustic wave, electron-acoustic wave, lower hybrid wave and upper hybrid wave. Chaotic, quasiperiodic and coexisting features are also highlighted in conservative and dissipative plasma systems.

Contents

Chapter 1

Introduction

1.1 Plasma as a state

In general, to a common person, there are three states of matter: solid, liquid and gas. Through exchange of energy, any one of the states can be converted to another state. For example, the H_2O molecule is a simple example, as it is available and seen in all these three states: ice (solid), water (liquid) and steam (gas). If we supply energy to matter which is in a gaseous state, it breaks the molecules into its component atoms and these atoms are stripped off their negative (electrons) and positive (ions) charged particles. The supplied energy can occur due to several causes, like heat, collision, reaction, and radiation, etc. This state of matter in which charged particles as well as neutral particles survive at the same time is normally known as plasma. The word "plasma" is a Greek word and it means something fabricated or molded. This word "plasma" was first utilized by Langmuir and Tonks in 1929 during their experiment. The plasma can be considered as a special phase as it has many distinguished properties compared to gaseous phase. Thus, the plasma [1] can be defined as:

Plasma is a quasineutral gas of charged and neutral particles which exhibits a collective behavior.

Since the plasma is observed naturally in the upper atmosphere of the earth, it is sometimes defined as

the fourth state of matter.

1.2 Plasmas exist in nature

Plasma does not exist on the earth naturally, but it can exist in some cosmic objects due to huge differences in temperature and density in the two regions. Planetary nebulae, hot stars, interstellar medium and upper atmosphere of the earth are some of the cosmic objects. Plasma can be produced in laboratories. Some regions, where natural plasmas are available, are described below.

1.2.1 Ionosphere

The earth's ionosphere is a part of the upper atmosphere which is partially ionized by solar radiation. It extends from 60 km to 1000 km altitude above the surface of Earth. The region contains a portion of the mesosphere, thermosphere and exosphere of earth's atmosphere. The temperature of the ionosphere typically lies between 300 to 1600 K and plasma density of the ionized particles ranges from 10^3 to 10^6 cm^{-3} [2]. The high temperature is due to the absorption of energized solar photons with other intense solar radiations like UV rays, X-rays and γ-rays. The solar stream ionizes the neutral particles of the atmosphere and produces equal numbers of electrons and positive ions. The free electrons of the ionosphere with sufficient density have a considerable effect on the propagation of radio frequency electromagnetic waves and, thus, the region is important for communication purposes.

1.2.2 Van Allen belts

Van Allen belt is a strata of highly energized particles held by the magnetic field of Earth. The belt is divided into two sub belts. One belt is extended roughly from 1000–6000 km and another ranges from 15000 km to 25000 km above the surface of the earth. The outer belt is created by the energetic and charged electrons whereas the inner belt contains protons and electrons. The other nuclei elements like alpha and beta particles also exist in smaller quantities.

1.2.3 Aurorae

The auroral zone is a ring of light emission created by the precipitation of atmospheric particles which are centered around the magnetic pole. Auroras are formed by the interaction of highly energized solar wind particles, like, electrons and protons ($T \sim 100$ eV) and ionospheric particles ($T \sim 1$ eV). Moreover, the Viking satellite reveals the presence of two-temperature electrons where the cold electron temperature must remain in the range $0.8 - 2.3$ eV and the hot electron temperature between 5 eV to 35 eV [3]. Generally, auroral emissions typically occur at altitudes of about 100 km, however, they may exist over 250 km above the Earth's surface. The solar wind particles captured by geomagnetic field are directed towards the magnetic poles and then collide with oxygen and nitrogen atoms to release highly excited ions with radiation at various wavelengths. Thus, this interaction creates the emission of lights with characteristic colors, such as, red or greenish blue. Mostly the greenish blue auroral emission is produced by ionized oxygen atoms at an altitude of 100–250 km and glow of red color is due to ionization of molecular oxygen at higher altitudes, up to 500 km. In addition to Earth, the atmospheres of planets like, Jupiter, Saturn, Uranus, and Neptune also exhibit the auroral activity.

1.2.4 Solar corona

Solar corona is the outermost atmosphere of the Sun. These layers extend thousands of kilometers from the solar surface and can be easily observable during a total solar eclipse. It has a temperature of several million degrees kelvin and the density is 10 billion times less than the Earth's atmosphere. Due to open magnetic flux, the outer layers of the solar corona constantly emit energized particles into the interplanetary space in the form of solar winds which flow at supersonic speed. In a magnetic medium, three types of waves that propagate into the interplanetary spaces emerge from the solar corona. These wave modes are the slow, fast, and the Alfvén wave modes. The slow and fast wave modes are compressive and grow into shock waves, on the contrary, the Alfvén wave is incompressive in nature [4].

1.2.5 Core of the sun

The core of the Sun or solar core is the innermost part and is considered as the hottest part of the Sun. It is about 0.2 to 0.25 of solar radius from the center and has a temperature of about 15 million degrees kelvin and a density of approximately 150 g/cm^3 at the center. The process of nuclear fusion takes place in the solar core where the combination of hydrogen nuclei results into the formation of helium. In addition to these heavy nuclei, the energy is released in the form of radiations and visible light.

1.2.6 HII regions

HII region is an astrophysical object which is mainly composed of a cloud of ionized hydrogen and is surrounded by high-energy UV radiations. This region crucially contributes to the formation of stars through clouds of molecular hydrogen and has a temperature of around 10,000 Kelvin. Their densities range from few atoms/cm^3, to millions of atoms/cm^3. The giant molecular clouds of hydrogen present in the HII region are embedded with weak magnetic fields with strengths of several nanoteslas.

1.3 Concept of temperature

To define "temperature", let us first consider a gas comprising of neutral or charged particles. These particles constantly collide with each other and travel at different velocities. They mostly move with Maxwell-Boltzmann (MB) distributed velocities. MB distribution is three-dimensional distribution but here one-dimensional MB distribution is considered as:

$$f(u) = Ae^{-\frac{1}{2}\frac{mu^2}{KT}}, \tag{1.1}$$

where u denotes velocity of the particle, $f(u)du$ denotes number of particles per unit length having velocities in the range from u to $u+du$, m denotes mass of a particle, K denotes the Boltzmann constant and T is the temperature of the gas. The number density of particles per unit length is denoted by n, where

$$
\begin{aligned}
n &= \int_{-\infty}^{\infty} f(u)du, \\
&= \int_{-\infty}^{\infty} Ae^{-\frac{1}{2}\frac{mu^2}{KT}} \, du.
\end{aligned}
$$

Defining $v_{th} = \sqrt{2KT/m}$ and $y = u/v_{th}$, we have

$$
\begin{aligned}
n &= A \int_{-\infty}^{\infty} e^{-y^2} v_{th} dy, \\
&= A v_{th} \int_{-\infty}^{\infty} e^{-y^2} dy, \\
&= 2 A v_{th} \int_{0}^{\infty} e^{-y^2} dy, \quad \text{[As the integrand is an even function]} \\
&= A v_{th} \int_{0}^{\infty} e^{-z} z^{\frac{1}{2}-1} dz, \quad \text{[Substituting } y^2 = z] \\
&= A v_{th} \Gamma\left(\frac{1}{2}\right), \quad \text{[Since } \Gamma(n) = \int_{0}^{\infty} e^{-z} z^{n-1} dz] \\
&= A v_{th} \sqrt{\pi}, \quad \text{[Since } \Gamma\left(\frac{1}{2}\right) = \sqrt{\pi}] \\
&= A \sqrt{\frac{2\pi KT}{m}}.
\end{aligned}
$$

Here, A is considered independent of particle velocity. Thus,

$$
A = n \sqrt{\frac{m}{2\pi KT}}.
$$

We see that A relies on temperature of the gas, hence, the distribution function (1.1) can be written as follows:

$$
f(u) = n \sqrt{\frac{m}{2\pi KT}} e^{-\frac{1}{2}\frac{mu^2}{KT}}.
$$

Change in the distribution function $f(u)$ w.r.t. velocity u is shown in Figure 1.1 for different values of temperatures. It is seen from Figure 1.1 that on increasing temperature, the number of particles having average velocity (corresponding to $u = 0$) lowers while width of the graph heightens such that the entire area under the curve remains constant, giving particle number density.

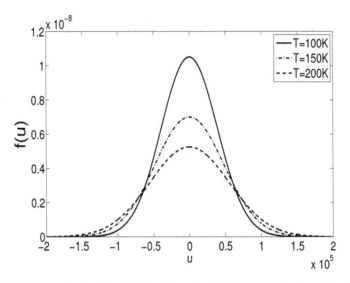

Figure 1.1: Change in the distribution function $f(u)$ w.r.t. velocity u for different values of temperature.

To define temperature T, one can calculate mean kinetic energy of the particles averaged over the MB-distribution (1.1) as

$$
\begin{aligned}
E_{av} &= \frac{\int_{-\infty}^{\infty} \frac{1}{2} mu^2 f(u) du}{\int_{-\infty}^{\infty} f(u) du}, \\
&= \frac{\int_{-\infty}^{\infty} \frac{1}{2} mu^2 A e^{-\frac{1}{2} \frac{mu^2}{KT}} du}{\int_{-\infty}^{\infty} A e^{-\frac{1}{2} \frac{mu^2}{KT}} du}.
\end{aligned}
\tag{1.2}
$$

Here, potential energy is neglected since the interaction between particles is presumed to be very feeble. Nevertheless, the particles collide briefly. Using the relations $v_{th} = \sqrt{2KT/m}$ and $y = u/v_{th}$ in equation (1.2), one can obtain

$$E_{av} = \frac{\frac{1}{2}mv_{th}^2 \int_{-\infty}^{\infty} y^2 e^{-y^2} dy}{\int_{-\infty}^{\infty} e^{-y^2} dy},$$

$$= \frac{\frac{1}{2}mv_{th}^2 \int_{0}^{\infty} y^2 e^{-y^2} dy}{\int_{0}^{\infty} e^{-y^2} dy}, \quad \text{[since the integrands are even functions]}$$

$$= \frac{\frac{1}{2}mv_{th}^2 \int_{0}^{\infty} e^{-z} z^{1/2} dz}{\int_{0}^{\infty} e^{-z} z^{1/2-1} dz}, \quad \text{[substituting } y^2 = z]$$

$$= \frac{\frac{1}{2}mv_{th}^2 \Gamma(3/2)}{\Gamma(1/2)}, \quad [\text{ since } \Gamma(n) = \int_{0}^{\infty} e^{-z} z^{n-1} dz]$$

$$= \frac{\frac{1}{2}mv_{th}^2 (\sqrt{\pi}/2)}{\sqrt{\pi}}, \quad \text{[since } \Gamma(3/2) = \frac{1}{2}\Gamma(1/2) \text{ and } \Gamma(1/2) = \sqrt{\pi}]$$

$$= \frac{1}{4}mv_{th}^2,$$

$$= \frac{1}{4}m\frac{2KT}{m},$$

$$= \frac{1}{2}KT. \tag{1.3}$$

Thus, for one-dimensional MB distribution, the average kinetic energy per degree of freedom is $KT/2$. As energy and temperature are related so closely, temperature is expressed in terms of units of energy. One eV energy is equivalent to a temperature of 11605 K.

1.3.1 *Existence of several temperatures*

It is important to point out that a plasma can have various kinetic temperatures at the same time. This is because ions and electrons can have independent Maxwell Boltzmann distributions corresponding to their different temperatures T_i and T_e, respectively. This can be explained as rates of collisions among the electrons and among the ions can be larger than those among the ions and electrons. As a result, the ions and electrons can have their independent thermal equilibriums. However, the situation of two different temperatures may not exist long and they equalize fast. It can be seen that if velocities of ions are much lower than the velocities of electrons, then temperature of the plasma is controlled by the electrons and so Maxwell Boltzmann distribution is given by the electron temperature T_e. If there is a magnetic field in a plasma, a single species can have two temperatures along the directions of magnetic field and normal to the magnetic field. These parallel and perpendicular components of velocities may have different Maxwell Boltzmann distributions corresponding to their temperatures along parallel and perpendicular directions.

1.3.2 Electron and ion temperatures

Plasmas can be produced in laboratories in a number of ways. One of the common methods is the electric discharge for producing low temperature plasma. To produce high temperature plasma in laboratories, one can use thermal ionization as temperature is the main cause for the production. A plasma composing of electrons, ions and neutrals provides a non-equilibrium property such that different components are not equally heated. In the case of an ordinary gas, all particles have the same mean kinetic energy of thermal motion. However, in the case of plasma, various plasma components have different mean kinetic energies. As a result, in a plasma, electrons have much higher energies than the ions and the kinetic energy of ions may be greater than that of the neutral particles. As mean kinetic energies of three components (electrons, ions and neutrals) are different, the plasma would be regarded as having three different temperatures T_e, T_i and T_n for the three species electrons, ions and neutrals, such that $T_e > T_i > T_n$.

1.3.3 Quasineutrality in plasma

Plasma is considered as electrically neutral at the macroscopic scale but the neutrality condition may not hold at the microscopic scale. Thermal energy associated with the plasma enables the movement of particles, resulting in local concentrations of electric charge. The electrical neutrality is restored due to presence of an electric field which is generated from any charge separation. At the microscopic scale, the balances between the thermal particle energy and electrostatic potential energy arising due to charge separation deviate plasma from maintaining the charge neutrality condition. However, at the macroscopic scale, this deviation does not occur, possibly due to the potential energy combined with the Coulomb forces that are comparatively greater than thermal kinetic energy of the particles. Thus, quasineutrality means that the plasma is neutral enough to maintain an equilibrium between the particles (i.e., $n_i \approx n_e$) but also preserve all interesting electromagnetic forces.

1.4   Debye length and Debye sphere

A chief characteristic of plasma is to shield out external electric potentials applied to it. To understand this, we consider the arrangement as given in Figure 1.2, i.e., we introduce an electric field in a plasma system by placing two charged spherical balls connected to an electric source, like a battery. We assume either shielding as a dielectric phenomena which restricts the plasma from recombination or an external source to be strong enough to balance the potential in the system. As electricity is supplied to the plasma system, the spherical charged balls start attracting the oppositely charged particles and clouds of positive ions and electrons from around the negative and positive balls, respectively.

If we consider cold plasma, then there is no thermal motions of charged particles. Thus, the clouds around positive and negative balls have an equal number of charges as the balls and there is no electric field outside the cloud. Consequently, the shielding is optimum. On the contrary, if the temperature of plasma is finite then the plasma particles lying on the boundary of the cloud can escape from the electrostatic potential well due to high thermal energy. We suppose that the boundary of the cloud occurs at the radial distance where the potential energy and kinetic energy (KT) of the particles are approximately equal. Therefore, shielding is not complete.

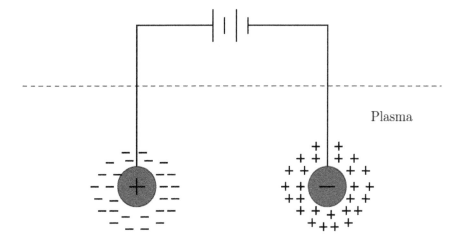

Figure 1.2: Two charged balls placed in a plasma and connected through an external electric source.

Now, a mathematical expression of the thickness of a charged cloud is to be obtained. Let the potential ϕ on the plane $x = 0$ with initial value ϕ_0. It is assumed that distance x is measured in radial direction. Now to evaluate $\phi(x)$, it is further assumed that ratio of the mass of positively charged ion and to that of electron M/m_e to be much higher, such that the positively charged ions are immobile and negatively charged electrons move freely in the charged gas. Radial variation is considered since condition is spherically symmetric. Thus, the one-dimensional form of the Poisson equation is given by

$$\varepsilon_0 \nabla^2 \phi = -q(n_i - n_e),$$
$$\text{or, } \varepsilon_0 \frac{d^2\phi}{dx^2} = -e(n_i - n_e), \tag{1.4}$$

where ε_0 is the absolute permittivity of the free space. Here, a simple plasma system is considered, i.e., hydrogen plasma, where the positively charged ions are protons. Now, density of protons is equal to that of the electrons (n) at large distance. Since protons are immobile, the density of protons n_i is n everywhere. Hence,

$$n_i = n. \tag{1.5}$$

In the presence of a potential energy $q\phi$ $(= -e\phi)$, the electron distribution function is

$$f(u) = A \, exp\left[-\left(\frac{1}{2}m_e u^2 - e\phi\right)/KT_e\right], \tag{1.6}$$

where T_e represents the temperature of the electron and u is the velocity of the electron. Now, electron density is

$$
\begin{aligned}
n_e &= \int_{-\infty}^{\infty} f(u) \, du \\
&= \int_{-\infty}^{\infty} A \, exp\left[-\left(\frac{1}{2}m_e u^2 - e\phi\right)/KT_e\right] du \\
&= A \, exp(e\phi/KT_e) \int_{-\infty}^{\infty} exp\left[-\left(\frac{1}{2}m_e u^2\right)/KT_e\right] du \\
&= 2A \, exp(e\phi/KT_e) \int_{0}^{\infty} exp\left[-\left(\frac{1}{2}m_e u^2\right)/KT_e\right] du. \tag{1.7}
\end{aligned}
$$

The following transformation is considered

$$\frac{1}{2}\frac{m_e u^2}{KT_e} = z,$$

$$\implies \frac{1}{2}u^2 = \frac{KT_e}{m_e}z. \tag{1.8}$$

When $u \to 0$, then $z \to 0$ and when $u \to \infty$, then $z \to \infty$. Now, differentiating equation (1.8) with respect to z, one can obtain

$$u\frac{du}{dz} = \frac{KT_e}{m_e}$$

$$\text{or, } du = \frac{1}{\sqrt{2}}\sqrt{\frac{KT_e}{m_e}}\, z^{\frac{1}{2}-1}\, dz. \tag{1.9}$$

Using equation (1.9) in equation (1.7), one can obtain

$$
\begin{aligned}
n_e &= 2A\, exp(e\phi/KT_e) \int_0^\infty \frac{1}{\sqrt{2}}\sqrt{\frac{KT_e}{m_e}}e^{-z}z^{\frac{1}{2}-1}\, dz, \\
&= 2A\, exp(e\phi/KT_e)\frac{1}{\sqrt{2}}\sqrt{\frac{KT_e}{m_e}} \int_0^\infty e^{-z}z^{\frac{1}{2}-1}\, dz, \\
&= A\, exp(e\phi/KT_e)\sqrt{\frac{2KT_e}{m_e}}\,\Gamma(1/2), \quad \left(\text{Since } \Gamma(n) = \int_0^\infty e^{-z}z^{n-1}\, dz\right), \\
&= A\, exp(e\phi/KT_e)\sqrt{\frac{2\pi KT_e}{m_e}}, \quad (\text{Since } \Gamma(1/2) = \sqrt{\pi}). \tag{1.10}
\end{aligned}
$$

At large distance, the boundary conditions are

$$x \to \infty, \qquad \phi \to 0, \qquad n_e \to n, \qquad n = A\sqrt{\frac{2\pi KT_e}{m_e}}. \tag{1.11}$$

Thus, using the boundary conditions, equation (1.10) becomes

$$n_e = n\, exp(e\phi/KT_e). \tag{1.12}$$

Now, substituting equations (1.5), (1.6) and (1.12) in (1.4), one can obtain

$$
\begin{aligned}
\varepsilon_0 \frac{d^2\phi}{dx^2} &= -e\,[n - n\,exp(e\phi/KT_e)] \\
&= en\,[exp(e\phi/KT_e) - 1] \\
&= en\,[1 + \frac{e\phi}{KT_e} + \frac{1}{2!}\left(+\frac{e\phi}{KT_e}\right)^2 + \cdots - 1] \\
&= en\,[\frac{e\phi}{KT_e} + \frac{1}{2!}\left(+\frac{e\phi}{KT_e}\right)^2 + \cdots]
\end{aligned}
$$

Thus,

$$
\varepsilon_0 \frac{d^2\phi}{dx^2} = \frac{ne^2}{KT_e}\phi \quad \text{[Neglecting second and higher degree terms of } \phi\text{]},
$$

$$
\implies \frac{d^2\phi}{dx^2} = \frac{\phi}{\lambda_D^2}, \tag{1.13}
$$

where

$$
\lambda_D = \left(\frac{\varepsilon_0 KT_e}{ne^2}\right)^{1/2}. \tag{1.14}
$$

which is a mathematical expression for Debye length. The sphere of radius λ_D is known as the Debye sphere.

1.5 Criteria for plasma

If ω is the frequency of plasma oscillation, L is linear dimension, λ_D is Debye length, N_D is Debye sphere, τ is mean time between collisions with neutral particles, then the criteria for an ionized gas to be called plasma are as follows:

i. linear dimension (L) of the plasma is much greater than the Debye length (λ_D), i.e., $\lambda_D << L$,

ii. number of plasma particles in Debye sphere (N_D) must be very much greater than unity, i.e., $N_D >> 1$,

iii. period of typical plasma oscillation must be much smaller than the mean time (τ) between collisions with neutral particles, i.e., $\frac{1}{\omega} < \tau$, i.e., $\omega\tau > 1$.

1.6 Plasma frequency

In a plasma, the equilibrium state is attained when plasma charged particles are evenly distributed such that the neutrality condition is satisfied everywhere. An electric field is generated when electrons are moved comparative to background ions. This creates an electric field in a direction that it attempts to draw the electrons in order to maintain the neutrality condition. Due to inertia, the electrons overshoot and an electric field is formed in an opposite direction that tends to draw electrons back to equilibrium position. Hence, the characteristic frequency at which the electrons oscillate at the equilibrium positions is defined as the plasma frequency. The oscillations of lower mass electrons are comparatively faster than that of the massive ion. This causes the ion particles to be unresponsive to the oscillating field created by the oscillations of electrons. Thus, the ions may be considered as fixed.

To obtain an expression for the plasma frequency denoted by ω_p, the following assumptions are considered:

i. The plasma filed has large area.

ii. There is no magnetic effect, so, from the Maxwell equations, one can obtain

$$\vec{\nabla} \times \vec{E} = -\dot{\vec{B}} = 0, \quad \text{and hence,} \quad \vec{E} = -\vec{\nabla}\phi,$$

where ϕ is scalar quantity.

iii. The ions being fixed form a uniform background in space.

iv. There are no thermal motions.

v. The electric field formed in the plasma results in the movement of electrons along one direction (x-direction, say). Then, the following relations hold

$$\vec{\nabla} \equiv \frac{\partial}{\partial x}\hat{i} \quad \text{and} \quad \vec{E} = E\,\hat{i}.$$

The movement of electrons is governed by the equations of motion and continuity, such as

$$m_e n_e \left[\frac{\partial \vec{u}_e}{\partial t} + (\vec{u}_e . \vec{\nabla})\vec{u}_e \right] = -e n_e \vec{E}, \tag{1.15}$$

$$\frac{\partial n_e}{\partial t} + \vec{\nabla}.(n_e \vec{u}_e) = 0. \tag{1.16}$$

Here, collisionless, homogeneous and isotropic plasma is considered. Such electrons support high frequency oscillations and are developed as a result of diversion from neutrality. Therefore, the Poisson equation for one-dimensional motion is given by

$$\varepsilon_0 \frac{\partial E}{\partial x} = e(n_i - n_e), \tag{1.17}$$

where n_e and n_i are number densities of electrons and ions, respectively. The equations (1.15)-(1.17) are solved using the linearisation method for the amplitude of each oscillating quantity being low. Here, the terms with two or higher order of amplitude are ignored. Firstly, segregate the variable into two parts: (i) an equilibrium part and, (ii) the perturbation part. The equilibrium part arises due to oscillations and shows the state of plasma in the absence of oscillations, and is designated by a subscript 0 while, the perturbation part is generated by oscillations, given by a subscript 1:

$$\begin{cases} n_e = n_0 + n_1, \\ \vec{u}_e = \vec{u}_0 + \vec{u}_1, \\ \vec{E} = \vec{E}_0 + \vec{E}_1. \end{cases} \tag{1.18}$$

Plasma is homogeneous and at rest before displacement of electrons takes place, thus

$$\begin{cases} \vec{\nabla} n_0 = \vec{u}_0 = \vec{E}_0 = 0, \\ \frac{\partial n_0}{\partial t} = \frac{\partial \vec{u}_0}{\partial t} = 0. \end{cases} \tag{1.19}$$

Substituting equations (1.18)-(1.19) in equations (1.15)-(1.17) and on linearisation, one can obtain

$$\begin{cases} m_e \frac{\partial \vec{u}_1}{\partial t} = -e\vec{E}_1, \\ \frac{\partial n_1}{\partial t} + n_0 \vec{\nabla}.\vec{u}_1 = 0 \\ \varepsilon_0 \frac{\partial E_1}{\partial x} = -en_1. \end{cases} \tag{1.20}$$

For simplicity, a one-dimensional case is considered and the oscillating quantities posses sinusoidal feature as $e^{i(kx - \omega t)}$, where k is wave number and ω is frequency. Then, time derivative $(\frac{\partial}{\partial t})$ is replaced by $-i\omega$, and $\vec{\nabla}$ by ik. From equation (1.20), one can obtain

$$\begin{cases} -i\omega m_e u_1 = -eE_1, \\ -i\omega n_1 = -n_0 i k u_1, \\ ik\varepsilon_0 E_1 = -en_1. \end{cases} \tag{1.21}$$

Now, from equation (1.21), one can obtain

$$-i\omega m_e u_1 = -i\frac{n_0 e^2}{\varepsilon_0 \omega} u_1, \tag{1.22}$$

such that

$$(m_e \omega - \frac{n_0 e^2}{\varepsilon_0 \omega}) u_1 = 0.$$

As u_1 is non-zero, one can obtain

$$\omega^2 = \frac{n_0 e^2}{m_e \varepsilon_0}.$$

Hence, the plasma frequency ω_p is defined as

$$\omega_p = \sqrt{\frac{n_0 e^2}{m_e \varepsilon_0}} \text{ rad/sec.} \tag{1.23}$$

Plasma frequency is also a basic parameter that depends on the plasma density. It is very high for electrons having lower mass.

1.7 Applications of plasma

Number density and temperature are two key parameters for which plasma has a collection of wide applications. Number density n can vary from 10^6 to 10^{34} m^{-3}. Similarly KT_e can vary from 0.1 to 10^6 eV. Most of the astrophysical bodies are in plasma state. Based on number densities and temperatures, there are several applications of plasmas. Some of them are discussed below.

1.7.1 Space physics

One of the important applications of plasma is the investigation of earth's environment in space, including the ionosphere and the magnetosphere. The nonlinear propagation of radio waves through the ionosphere, creation of Van Allen radian belts and aurora borealis can be realized using the study of plasma dynamics.

1.7.2 Astrophysics

It is important to note that most of the astrophysical objects are in plasma state. Basics of plasma theories are being used to describe various astrophysical phenomena with the expectation of de-mystifying the origin of the universe.

1.7.3 Gas lasers

The evolution of high power lasers in the 1960s directed a new area of research which is known as laser plasma. A significant application of laser plasma is to survey the prospect of inertial confinement fusion. The strong electric field obtained when a high power laser pulse goes through a plasma is anticipated to be used for particle acceleration. This approach can dramatically minimize the size and cost of conventional particle accelerators. The laser produced plasmas have utmost conditions of density and temperature which are not observed in classical plasmas. For high density and low temperature plasmas, quantum effects play a significant role and it is required to develop quantum plasma—a new direction of plasma research.

1.7.4 Industrial application

There are several industrial applications of plasmas. For example:

 i. Plasma light, Plasma TVs,

 ii. Fabrication, thin film processing,

 iii. Biomedical and medicine,

 iv. Synthesis,

 v. Surface maintenance,

 vi. Laser.

1.8 Fluid description of plasma

A plasma is a collection of large number of charged and neutral particles. For this reason, it is not possible to identify the trajectory of each particle to predict the behaviour of plasma. It is seen that a typical plasma density is of order 10^{12} ion-electrons pairs per cm^3. It is observed that 80% of the plasma phenomena can be explained by fluid model. In this situation, one can consider motion of fluid elements instead of motion of individual particles. In this book, we shall concentrate only on fluid approach to study nonlinear waves in plasmas. A more defined approach, which is known as the kinetic theory approach, demands more mathematical calculation and is not under consideration for this book.

1.8.1 Maxwell's equation

Maxwell's equation [1] narrates how electric and magnetic fields are generated by charges, currents and changes of each other. The following four fundamental equations are known as the Maxwell's equation:

i. $\vec{\nabla} \cdot \vec{D} = \rho$, or $\vec{\nabla} \cdot \vec{E} = \frac{\rho}{\varepsilon_0}$,

ii. $\vec{\nabla} \cdot \vec{B} = 0$,

iii. $\vec{\nabla} \times \vec{E} = -\frac{\partial \vec{B}}{\partial t}$,

iv. $\vec{\nabla} \times \vec{H} = \vec{J} + \frac{\partial \vec{D}}{\partial t}$,

where ε_0 = permittivity of the free space,

$\vec{E}$ = Electric field,
$\vec{D}$ = Electric displacement vector which is $\vec{D} = \varepsilon_0 \vec{E} + \vec{p}$,
p = pressure,
ρ = charge density,
$\vec{B}$ = Magnetic induction,
$\vec{H}$ = Magnetic field intensity, and
$\vec{J}$ = current density.

1.8.2 Equation of motion

From Maxwell's equations, one can understand the effects of the electric field $\vec{E}$ and the magnetic field $\vec{B}$ for the given plasma state.

A plasma particle of mass m with charge q is considered to be moving in $\vec{E}$ and $\vec{B}$ with velocity $\vec{v}$. Then, the corresponding Lorentz force acting on the particle is $q[\vec{E} + \vec{v} \times \vec{B}]$. Then, the equation of motion [5] of the plasma particle is expressed as

$$m\frac{d\vec{v}}{dt} = q[\vec{E} + \vec{v} \times \vec{B}]. \tag{1.24}$$

It is considered that there is no collision and thermal motion of the plasma particles. Then, all the particles in the fluid element move together and their average velocity $\vec{u}$ is the same as the velocity of the single particle.

So, the equation of motion for n plasma particles, i.e., for a fluid element is

$$mn\frac{d\vec{u}}{dt} = qn[\vec{E} + \vec{u} \times \vec{B}]. \tag{1.25}$$

Now one can write

$$
\begin{aligned}
\frac{d}{dt}\vec{u}(x,y,z,t) &= \frac{\partial \vec{u}}{\partial t} + \frac{\partial \vec{u}}{\partial x}\frac{dx}{dt} + \frac{\partial \vec{u}}{\partial y}\frac{dy}{dt} + \frac{\partial \vec{u}}{\partial z}\frac{dz}{dt}, \\
&= \frac{\partial \vec{u}}{\partial t} + \frac{\partial \vec{u}}{\partial x}u_x + \frac{\partial \vec{u}}{\partial y}u_y + \frac{\partial \vec{u}}{\partial z}u_z, \\
&= \frac{\partial \vec{u}}{\partial t} + (u_x\hat{i} + u_y\hat{j} + u_z\hat{k}) \cdot \left(\frac{\partial}{\partial x}\hat{i} + \frac{\partial}{\partial y}\hat{j} + \frac{\partial}{\partial z}\hat{k}\right)\vec{u}, \\
&= \frac{\partial \vec{u}}{\partial t} + (\vec{u} \cdot \vec{\nabla})\vec{u}. \tag{1.26}
\end{aligned}
$$

Using equation (1.26) in equation (1.25), one can obtain

$$mn\left[\frac{\partial \vec{u}}{\partial t} + (\vec{u} \cdot \vec{\nabla})\vec{u}\right] = qn[\vec{E} + \vec{u} \times \vec{B}]. \tag{1.27}$$

If the thermal motion of the particles is considered, then one pressure term has to be added in the equation of motion (1.26). Then, equation (1.27) becomes

$$mn\left[\frac{\partial \vec{u}}{\partial t} + (\vec{u} \cdot \vec{\nabla})\vec{u}\right] = qn[\vec{E} + \vec{u} \times \vec{B}] - \vec{\nabla}p, \tag{1.28}$$

where p is the scalar kinetic pressure.

If collision between neutral particles is considered with velocity $\vec{u}_0$ and charged particles with velocity $\vec{u}$, the resulting force due to collision is $-mn\frac{\vec{u}-\vec{u}_0}{\tau}$, where τ is the mean free time between the collisions. Then, equation (1.28) becomes

$$mn\left[\frac{\partial \vec{u}}{\partial t} + (\vec{u} \cdot \vec{\nabla})\vec{u}\right] = qn[\vec{E} + \vec{u} \times \vec{B}] - \vec{\nabla}p - mn\frac{\vec{u}-\vec{u}_0}{\tau}. \tag{1.29}$$

For any anisotropic fluid $\vec{\nabla}p$ in equation (1.29) is replaced by $\vec{\nabla}P$, where P is the stress tensor.

Then equation (1.28) takes the form

$$mn\left[\frac{\partial \vec{u}}{\partial t} + (\vec{u} \cdot \vec{\nabla})\vec{u}\right] = qn[\vec{E} + \vec{u} \times \vec{B}] - \vec{\nabla}P. \tag{1.30}$$

In absence of $\vec{E}$ and $\vec{B}$, equation (1.30) becomes

$$mn\left[\frac{\partial \vec{u}}{\partial t} + (\vec{u} \cdot \vec{\nabla})\vec{u}\right] = -\vec{\nabla}P. \tag{1.31}$$

In case of an ordinary fluid, collisions among its constituent particles are quite frequent and the fluid follows the Navier-Stokes equation as

$$\rho\left[\frac{\partial \vec{u}}{\partial t} + (\vec{u} \cdot \vec{\nabla})\vec{u}\right] = -\vec{\nabla}p + \rho\gamma\nabla^2 \vec{u}, \tag{1.32}$$

where, ρ is the mass density and γ is the coefficient of Kinematic viscosity. Now $mn = \rho$ is the mass density and if $\rho\gamma\nabla^2 \vec{u} = \vec{\nabla}p - \vec{\nabla}P$ is considered, then equation (1.32) is reduced to equation (1.31). It is important to note that particles do not collide for equation (1.31), but collisions among the particles are quite frequent for equation (1.32).

It is important to note that the fluid model works for plasma can be provided in terms of the role played by the magnetic field as that of collisions, in a special sense. If we use magnetic field $\vec{B}$ in addition to electric field $\vec{E}$, then $\vec{B}$ also limits free streaming of the particles by turning them to gyrate in the Larmor orbits. In this way, a collisionless plasma acts like a collisional fluid in presence of magnetic field $\vec{B}$. Hence, a fluid theory can be taken as a valid theory for plasmas.

1.8.3 *Equation of continuity*

The equation of continuity [6] can be obtained from the principle of conservation of mass. From conservation of mass, it is known that the rate of decrease of number of plasma particles in a volume V inside a fluid plasma is equal to the number of particles that leave the volume V per unit time through the surface S. Then, one can obtain

$$-\frac{\partial}{\partial t}\oint_V n\, dV = \oint_S n\vec{u} \cdot d\vec{S}, \tag{1.33}$$

where $\vec{u}$ denotes the velocity of the fluid plasma and n is the number density of plasma particle.

From Gauss's law of divergence theorem, it is known to us

$$\oint_S n\vec{u} \cdot d\vec{S} = \oint_V \vec{\nabla} \cdot (n\vec{u}) \, dV. \tag{1.34}$$

From equations (1.33) and (1.34), it can be written as

$$-\frac{\partial}{\partial t} \oint_V n \, dV \;=\; \oint_V \vec{\nabla} \cdot (n\vec{u}) \, dV,$$

$$\text{or,} \quad \oint_V \frac{\partial n}{\partial t} \, dV \;=\; -\oint_V \vec{\nabla} \cdot (n\vec{u}) \, dV. \tag{1.35}$$

Since both integrals are for the same volume V, their integrands must be equal. So, the following relation is always true

$$\frac{\partial n}{\partial t} = -\vec{\nabla} \cdot (n\vec{u}),$$

$$\text{or,} \quad \frac{\partial n}{\partial t} + \vec{\nabla} \cdot (n\vec{u}) = 0 \tag{1.36}$$

which is familiar as the equation of continuity for each type of plasma particles.

1.8.4 Equation of state of plasma

In a fluid plasma [1], where the change of energy from outside of plasma is not allowed, the pressure p and density ρ of the fluid are related to

$$p = C\rho^{\gamma}, \tag{1.37}$$

where C is a constant,
$\quad \gamma = \frac{C_p}{C_v}$ = ratio of specific heat,
$\quad C_p$ = specific heat under constant pressure,
$\quad C_v$ = specific heat under constant volume.

So, equation (1.37) becomes

$$p \;=\; C(mn)^{\gamma}, \tag{1.38}$$
$$\;=\; Cm^{\gamma}n^{\gamma},$$
$$\;=\; C_1 n^{\gamma}, \tag{1.39}$$

where $C_1(= Cm^\gamma)$ is a constant, m denotes the mass of plasma particles and n denotes number density of the plasma particles.

For a plasma system with N degrees of freedom, the following relation is true

$$\gamma = 1 + \frac{2}{N}. \tag{1.40}$$

1.8.5 Poisson equation

If the given plasma system consists of electrons and ions with number densities n_e and n_i, respectively, then the plasma system is closed by the Poisson equation as

$$\varepsilon_0 \, \vec{\nabla} \cdot \vec{E} = e(n_i - n_e),$$
$$\text{or } \varepsilon_0 \, \vec{\nabla} \cdot (-\vec{\nabla}\phi) = e(n_i - n_e),$$
$$\text{or } \varepsilon_0 \, \vec{\nabla}^2 \phi = e(n_e - n_i), \tag{1.41}$$

where $\vec{E} = -\vec{\nabla}\phi$, ϕ denotes electrostatic potential, e is elementary charge and ε_0 is permittivity of free space.

References

[1] F. F. Chen, Introduction to Plasma Physics and Control Fusion, Springer, Newyork (2008).

[2] M. Kelly, The Earth's ionosphere: Plasma physics and electrodynamics. Vol. 43. Elsevier (2012).

[3] M. Berthomier, R. Pottelette and M. Malingre, Journal of Geophysical Research: Space Physics, 103(A3): 4261–4270 (1998).

[4] G. K. Parks, Physics of Space Plasmas. An Introduction, University of California, Berkley (1984).

[5] J. A. Bittencourt, Fundamentals of Plasma Physics, Springer-Verlag, Newyork (2004).

[6] B. Ghosh, Basic Plasma Physics, Alpha Science International (2014).

Chapter 2

Dynamical Systems

2.1 Introduction to dynamical systems

This chapter deals with systems that undergo change and show significant behaviors as they evolve with time. The main focus here is to examine motions of particles in dynamical systems [1]. The dynamical systems are divided mainly into two types: differential equations and iterated maps. The evolution of dynamical systems in continuous time is described by differential equations, while for the systems in which time is discrete the iterated maps are applied. The time-dependent dynamical systems are known as nonautonomous dynamical systems and time-independent dynamical systems are called autonomous dynamical systems. In one-dimensional spaces, the motions of particles in dynamical systems are highly restricted to a definite region or remain constant. However, in higher dimensional spaces, the motions of particles are free to move in any region and, hence, give rise to interesting behaviors. The wide range of dynamical features of a system can be examined for second-order dimensional systems.

2.1.1 One-dimensional system

An one-dimensional dynamical system can be represented as

$$\dot{x} = f(x), \tag{2.1}$$

where $x(t)$ and $f(x)$ are real valued functions of time t and space variable x, respectively. The system (2.1) represents a vector field with velocity vector $\dot{x}$ at each point x on the x-axis. To show the vector field, one can plot $\dot{x}$ versus

x, and put arrows on the x-axis to present the velocity vector $\dot{x}$ at each point x. If $\dot{x} > 0$, one can put an arrow directed to the right and to the left for $\dot{x} < 0$.

2.1.1.1 Equilibrium point and its stability

The motion of the system (2.1) is on the right for $\dot{x} > 0$ and on the left for $\dot{x} < 0$. A point for which $\dot{x} = 0$, i.e., there is no motion, is called a fixed point or equilibrium point of the system (2.1). A fixed point is said to be stable if the motion is towards it and unstable if the motion is away from it. A stable fixed point is also called attractor or sink. Similarly, an unstable fixed point is called repeller or source [2].

2.1.1.2 Trajectory and phase portrait

In a dynamical system, one can consider a particle moving in a space from coordinates $(x_1(0), x_2(0))$ to $(x_1(t), x_2(t))$, where 0 refers to the initial position and t refers to its final position. Then, the solution of the system corresponds to curve traced by the particle moving in that space. The above-mentioned curve is known as trajectory and the corresponding space is known as phase space [3]. Each point of trajectory in phase space can be considered as an initial condition. Such trajectory in phase space provides information about the solution.

2.1.1.3 Example

Find all fixed points of the system $\dot{x} = x^2 - 4$ and discuss their stabilities.

Solution: Consider $f(x) = x^2 - 4$ and set $f(x^*) = 0$, where x^* denotes fixed point. Then, one can get fixed points $x^* = \pm 2$. In order to examine the stability, a curve is plotted for $f(x) = x^2 - 4$ in the vector field $(x, \dot{x})$-plane. For $f(x) = x^2 - 4 > 0$, the flow is towards the right side and to the left side when $f(x) = x^2 - 4 < 0$. Therefore, $x^* = -2$ is a stable fixed point and $x^* = 2$ is an unstable fixed point (see Figure 2.1).

2.1.2 Linear stability analysis

A graphical method is an effective tool to analyze the stability of systems. The rate at which the system grows or decays to a stable fixed point can be easily interpreted from the graphical method. This information can be acquired by linearizing about a fixed point. A small perturbation

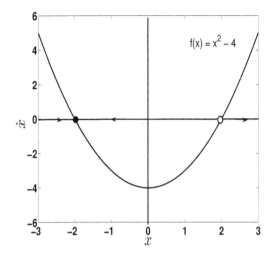

Figure 2.1: Graph for x vs. $\dot{x}$ for Example 2.1.1.3.

$\zeta(t) = x(t) - x^*$ can be introduced away from x^*, where x^* is a fixed point. To examine the behavior of perturbation, one can obtain an ordinary differential equation for ζ which gives

$$\dot{\zeta} = \frac{d}{dt}(x - x^*) = \dot{x},$$

where x^* is constant. Therefore, $\dot{\zeta} = \dot{x} = f(x) = f(x^* + \zeta)$. Applying the Taylor's series one can get

$$f(x^* + \zeta) = f(x^*) + \zeta f'(x^*) + O(\zeta^2),$$

here $O(\zeta^2)$ is quadratically lesser values in ζ. Lastly, $f(x^*) = 0$ at fixed point x^*. Thus,

$$\dot{\zeta} = \zeta f'(x^*) + O(\zeta^2).$$

For $f'(x^*) \neq 0$, the terms of $O(\zeta^2)$ become negligible and can be approximated as

$$\dot{\zeta} \approx \zeta f'(x^*).$$

This process is called linearization in ζ about x^*. It indicates that the perturbation $\zeta(t)$ increases exponentially for $f'(x^*) > 0$ and diminishes for $f'(x^*) < 0$. If $f'(x^*) = 0$, the $O(\zeta^2)$ terms are not negligible and a nonlinear analysis is needed to determine stability. The outcome is that the slope $f'(x^*)$ at the fixed point determines its stability.

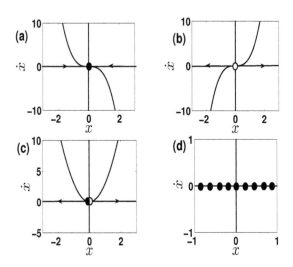

Figure 2.2: Phase plots for systems (a)–(d) in Example 2.1.2.1.

2.1.2.1 Example

Discuss the stability of fixed points for the following systems:
(a) $\dot{x} = -2x^3$, (b) $\dot{x} = 2x^3$, (c) $\dot{x} = -3x^2$ (d) $\dot{x} = 0$.

Solution: The stability is determined by graphical method. Here, fixed point x^* is 0 for all the systems (a)-(d). However, there are different types of stabilities of the fixed points. Figure 2.2 depicts that (a) the fixed point is stable and (b) the fixed point is unstable. Here, for the system (c) the fixed point is half-stable whereas, in case (d), all points of the whole line (*x*-axis) are fixed points.

2.1.3 Potentials

The first-order nonlinear dynamical system $\dot{x} = f(x)$ can be determined using the physical concept of energy, preferably known as potential energy. A particle is sloping down potential well walls, where $V(x)$ is the potential defined by

$$f(x) = -\frac{dV}{dx}.$$

Consider a particle that is highly damped. The damping of a particle means that its inertia is insignificant with respect to the force due to damping and potential. The negative sign in the expression of V indicates downward motion of the particles. To examine this one can consider x as a function of t. Then, differentiating $V(x(t))$ w.r. to t, one can get

$$\frac{dV}{dt} = \frac{dV}{dx}\frac{dx}{dt}.$$

Since $\dot{x} = f(x) = -\frac{dV}{dx}$, one can obtain

$$\frac{dx}{dt} = -\frac{dV}{dx},$$

and thus,

$$\frac{dV}{dt} = -\left(\frac{dV}{dx}\right)^2 \le 0.$$

This relation indicates that $V(t)$ gradually slows down along trajectories. The potential V of the particle becomes invariant at a fixed point as $dV/dx = 0$. It is also seen that a stable fixed point is attained at local minima of $V(x)$, while an unstable fixed point is attained at local maxima.

2.1.3.1 Example

Plot the potential for the system $\dot{x} = -4x$ and obtain all possible fixed points.

Solution: Consider $f(x) = -4x$. The only fixed point (x^*) of the system is 0. From the definition, one can have $-\frac{dV}{dx} = f(x) = -4x$. Then, potential function becomes $V(x) = 2x^2 + C$, where C is an integral constant. For simplicity, one can consider $C = 0$. From Figure 2.3, the plot of potential curve is shown, where the fixed point $x^* = 0$ is stable.

2.1.3.2 Example

Find all the fixed points and plot the potential for the system $\dot{x} = 2x - 4x^3$.

Solution: Consider $f(x) = 2x - 4x^3$ and x-coordinate of the fixed points of the system are 0 and $\pm\sqrt{\frac{1}{2}}$. Then, $-\frac{dV}{dx} = f(x) = 2x - 4x^3$, for which the potential becomes $V(x) = x^4 - x^2 + C$, where C is an integral constant. Then, one can set $C = 0$. From the plot of potential V shown in Figure 2.4, the fixed point $(0,0)$ is a local maximum that corresponds to an unstable point, while local minima at fixed points $(\pm\sqrt{\frac{1}{2}},0)$ are stable. The feature of potential

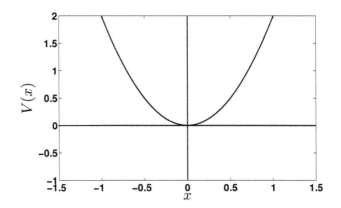

Figure 2.3: Plot of potential $V(x)$ vs. x for Example 2.1.3.1.

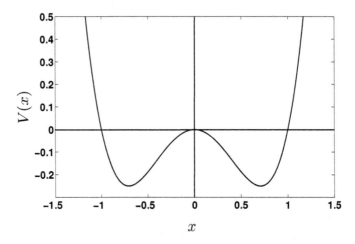

Figure 2.4: Plot of potential $V(x)$ vs. x for Example 2.1.3.2.

curve shown in Figure 2.4 is mostly known as a potential with double-well feature and the corresponding system is known as bistable as there exist two stable fixed points.

2.1.4 Bifurcations

In 1885, Henry Poincaré introduced the term "bifurcation". The qualitative structure of the motion can differ as parameters are varied. Specifically, fixed points can be generated or demolished or their stabilities can be modified.

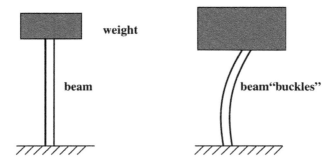

Figure 2.5: Buckling of beam.

These significant changes in the dynamics of the system are called bifurcations and the parameter values at which bifurcations occur are called bifurcation points [2].

Bifurcations describe systems that undergo transitions and instabilities due to variations in control parameters. Example given, a buckling of beam [2] shown in Figure 2.5. This figure depicts that if the weight given on the top of the beam is light then the beam can stand still in its position. But, when the weight is heavier on the beam then it becomes unstable and the beam may buckle. This phenomenon is called bifurcation, where deflection of the beam can be observed with weight being the control parameter.

There are different types of bifurcations:

1. Saddle node bifurcation,

2. Transcritical bifurcation and

3. Pitchfork bifurcation which is divided into two subcategories: super-critical and subcritical pitchfork bifurcations.

2.1.5 *Linear system in two-dimension*

Consider a two-dimensional linear system of the form

$$\begin{cases} \dot{x} = px + qy, \\ \dot{y} = rx + sy, \end{cases} \tag{2.2}$$

where p, q, r and s are parameters.

The system (2.2) can be written in a matrix form as

$$\dot{X} = AX,$$

where $A = \begin{bmatrix} p & q \\ r & s \end{bmatrix}$ and $X = \begin{bmatrix} x \\ y \end{bmatrix}$.

Therefore, the characteristic equation is

$$det(A - \lambda I) = 0$$

$$\text{i.e.,} \quad \begin{vmatrix} p - \lambda & q \\ r & s - \lambda \end{vmatrix} = 0.$$

Expanding the determinant yields

$$\lambda^2 - \tau\lambda + \Delta = 0, \tag{2.3}$$

where $\tau = \text{trace}(A) = p + s$ and $\Delta = \det(A) = ps - rq$.

Then

$$\lambda_1 = \frac{\tau + \sqrt{\tau^2 - 4\Delta}}{2}, \quad \lambda_2 = \frac{\tau - \sqrt{\tau^2 - 4\Delta}}{2} \tag{2.4}$$

are the solutions of the quadratic equation (2.3), where $\Delta = \lambda_1\lambda_2$, and $\tau = \lambda_1 + \lambda_2$.

Classifications are based on the values and signs of Δ and $\tau^2 - 4\Delta$ [2].

When $\tau^2 - 4\Delta > 0$ and $\Delta > 0$, the fixed point is a node for real eigenvalues having same sign. When $\tau^2 - 4\Delta < 0$ and $\Delta > 0$, the fixed points are either spirals or centers as the eigenvalues are complex conjugate. The curve $\tau^2 - 4\Delta = 0$ is a parabola which acts as a margin between spirals and nodes. Degenerate nodes and star nodes reside on this parabola. When $\Delta < 0$, the fixed point is a saddle for real eigenvalues having opposite signs. While the fixed point is a center for $\Delta > 0$, $\tau = 0$ and $\tau^2 - 4\Delta < 0$. The fixed points are stars/degenerate nodes for $\Delta > 0$ and $\tau^2 - 4\Delta = 0$.

The stability of spirals and nodes is determined from the values of τ. If $\tau < 0$, then it is stable, otherwise it is unstable. The fixed points can be classified on

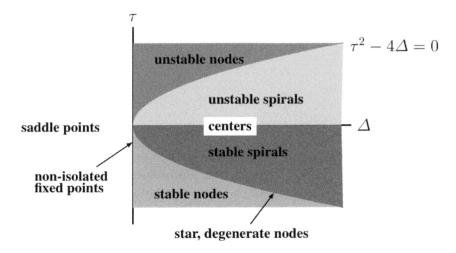

Figure 2.6: Classification of equilibrium points on the basis of Δ and τ.

the basis of eigenvalues. It is a stable node when λ_1 and λ_2 are both negative and an unstable node when both are positive. For real and repeated eigenvalues ($\lambda_1 = \lambda_2 = \lambda$ (say)), then the fixed point is stable star node if $\lambda < 0$ and unstable star node if $\lambda > 0$. For complex eigenvalues $(\alpha \pm i\beta), \alpha, \beta \in R$, unstable spiral is attained for $\alpha > 0$, stable spiral for $\alpha < 0$, and stable center for $\alpha = 0$.

2.1.5.1 *Example*

Draw phase portrait if eigenvalues are complex numbers.

Solution: As established in the above section, the fixed point is a center or spiral when eigenvalues are complex. When it is purely imaginary, the fixed points are centers. Center fixed points are developed by a class of closed orbits and are neutrally stable. Such closed orbits are not dragged by sink or source fixed points. The class of orbits enclosing a single fixed point that is the center implies the periodic solution of the system. A spiral orbit exists in a system if the system is slightly damped. The orbits about that fixed point are not in a closed form.

Consider the respective systems (S_1) and (S_2) as

$$\begin{cases} \dot{x} = y \\ \dot{y} = ax - bx^3, \end{cases} \tag{2.5}$$

$$\begin{cases} \dot{x} = y \\ \dot{y} = ax - bx^2 + cy, \end{cases} \tag{2.6}$$

where a, b, c are constants. Let the eigenvalues be $\lambda_{1,2} = \alpha \pm i\beta$, where $\alpha = \tau/2$ and $\beta = 1/2\sqrt{4\Delta - \tau^2}$. The complex eigenvalues are obtained if $\tau^2 - 4\Delta < 0$. Here, the system (S_1) has complex roots for $a < 0$ and $b > 0$. Thus, the fixed point is a center at $(0,0)$. In the case of S_2, the phase portrait 2.7(b) depicts spiral about the fixed point. It is a stable spiral if the real part of the complex eigenvalue is less than zero, whereas, it is an unstable spiral for the real part of the complex eigenvalue greater than zero.

2.1.6 Phase plane analysis

The solution of $\dot{X} = AX$ can be presented as a moving trajectory in the phase plane. The system can be structured as a vector field on the phase space [4]. The trajectory of a particle projects from one position to another showing the dynamical feature in a phase space. Thus, the phase portrait or phase space of the dynamical system presents the overall motion of trajectories. In Figure 2.7(a), trajectories start propagating in a circular motion, forming a small closed form. Such closed trajectories correspond to periodic solutions of the system. Therefore, any trajectory in a phase portrait provides informa-

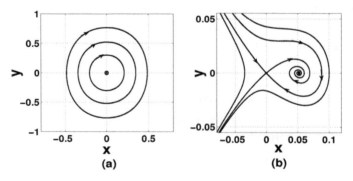

Figure 2.7: Phase portraits for complex eigenvalues where (a) center and (b) spiral.

tion of solution for the system. Figures 2.7(a) and (b) are the phase portraits of S_1 and S_2, respectively.

2.1.6.1 Nonlinear system in two-dimension

The nonlinear system is considered

$$\begin{cases} \dot{x} = f(x,y), \\ \dot{y} = g(x,y), \end{cases} \tag{2.7}$$

and assume that (x^*, y^*) is a fixed point.

Consider small perturbations about fixed points as $u = x - x^*$, $v = y - y^*$.

In order to examine the behavior of the perturbation, one can obtain differential equations as,

$$\dot{u} = u\frac{\partial f}{\partial x} + v\frac{\partial f}{\partial y} + O(u^2, v^2, uv)$$

and

$$\dot{v} = u\frac{\partial g}{\partial x} + v\frac{\partial g}{\partial y} + O(u^2, v^2, uv).$$

The perturbation (u, v) corresponds to

$$\begin{pmatrix} \dot{u} \\ \dot{v} \end{pmatrix} = \begin{pmatrix} \frac{\partial f}{\partial x} & \frac{\partial f}{\partial y} \\ \frac{\partial g}{\partial x} & \frac{\partial g}{\partial y} \end{pmatrix} \begin{pmatrix} u \\ v \end{pmatrix} + \text{quadratic terms.}$$

The matrix

$$A = \begin{pmatrix} \frac{\partial f}{\partial x} & \frac{\partial f}{\partial y} \\ \frac{\partial g}{\partial x} & \frac{\partial g}{\partial y} \end{pmatrix}_{(x^*, y^*)}$$

is the corresponding Jacobian matrix [1] at the equilibrium point (x^*, y^*) of the equation (2.7). Let $\Delta = \det A(x^*, y^*)$.

The theory of planar dynamical systems [3] describes that the fixed point (x^*, y^*) of the dynamical system (2.7) is a

■ saddle point if $\Delta < 0$ and trace $A = 0$,

■ center if $\Delta > 0$ and trace $A = 0$,

■ cusp if $\Delta = 0$ and trace $A = 0$.

2.1.6.2 Conservative system

The foundation of various significant second-order systems is Newton's law $F = ma$, where m is mass of a particle traveling in x-axis with nonlinear force $F(x)$ and a is acceleration of the particle. Then, the equation of motion is given by

$$m\ddot{x} = F(x).$$

Here, F is not dependent on $\dot{x}$ and t. To show that the energy is conserved, consider a potential energy function denoted by $V(x)$. Then, by definition of potential, one can have $F(x) = -dV/dx$. Then

$$m\ddot{x} + \frac{dV}{dx} = 0.$$

Multiplying both sides by $\dot{x}$, one can have

$$\frac{d}{dt}\left[\frac{1}{2}m\dot{x}^2 + V(x)\right] = 0.$$

Now using the chain rule, one can obtain

$$\frac{d}{dt}V(x) = \frac{dV}{dx}\frac{dx}{dt}.$$

Thus, the total energy

$$E = \frac{1}{2}m\dot{x}^2 + V(x)$$

remains invariant as a time function. Then, this energy E is a conserved value which is consistent with motion. The corresponding systems are known as conservative systems. Examples are given below.

2.1.6.3 Example

For $m = 1$, suppose a particle traveling with potential $V(x) = -x^2 + x^4$. Obtain all the fixed points and plot the phase portrait for the system.

Solution: From Example 2.1.3.2, one can have $-\frac{dV}{dx} = 2x - 4x^3$. The corresponding equation of motion is $\ddot{x} = 2x - 4x^3$ which can be written in the form of vector field

$$\dot{x} = y$$
$$\dot{y} = 2x - 4x^3.$$

Here, the velocity of particle is given by y. The fixed points of the system are $(0,0)$ and $(\pm\sqrt{\frac{1}{2}},0)$. Then the Jacobian matrix will be

$$A = \begin{pmatrix} 0 & 1 \\ 2 - 12x^2 & 0 \end{pmatrix}$$

Then, the determinant of $A = 12x^2 - 2$. The fixed point $(0,0)$ is a saddle since $\det A < 0$ and centers at $(\pm\sqrt{\frac{1}{2}},0)$ for $\det A > 0$. Thus, energy (E) of the system is $E = \frac{y^2}{2} - x^2 + x^4 = $ constant.

Figure 2.8 shows the system has a saddle at the origin and centers at $(\pm\sqrt{\frac{1}{2}},0)$. There exist fixed points that are enclosed by the family of small closed trajectories. Those trajectories correspond to periodic solutions of the system, while the larger closed trajectory corresponds to superperiodic solution of the system. The trajectory starting at fixed point $(0,0)$ and terminat-

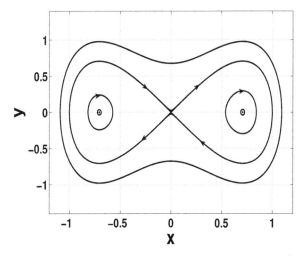

Figure 2.8: Phase portrait for system $\ddot{x} = 2x - 4x^3$.

ing at the same point is called homoclinic trajectory. Such trajectories occur mostly in conservative systems.

2.1.6.4 Example

Plot the phase portrait for the system $\ddot{x} = ax - bx^3$, considering constants $a, b < 0$.

Solution: The vector field form of $\ddot{x} = ax - bx^3$ is given by

$$\dot{x} = y$$

$$\dot{y} = ax - bx^3.$$

The fixed points of the systems are $(0,0)$ and $(\pm\sqrt{\frac{a}{b}}, 0)$. Therefore the Jacobian matrix is

$$A = \begin{pmatrix} 0 & 1 \\ a - 3bx^2 & 0 \end{pmatrix}$$

The determinant of $A = -a + 3bx^2$. Since $a, b < 0$, the fixed point at $(0,0)$ is center as det$A > 0$. However, the fixed points at $(\pm\sqrt{\frac{a}{b}}, 0)$ are saddle points as det$A < 0$. Thus, energy (E) of the system is $E = \frac{y^2}{2} - \frac{a}{2}x^2 + \frac{b}{4}x^4$.

Figure 2.9 shows that the fixed point is center at origin and saddle points at $(\pm\sqrt{\frac{a}{b}}, 0)$. The family of small closed trajectories correspond to periodic

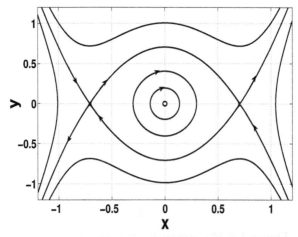

Figure 2.9: Phase portrait for $a, b < 0$ of the system $\ddot{x} = ax - bx^3$.

solutions of the system. The trajectory starts from one fixed point and ends at the other fixed point, this is called heteroclinic trajectory.

2.1.6.5 *Hamiltonian system*

The study of the Hamiltonian system $H(p,q)$ provides a fundamental and geometric edition of Newton's laws. Suppose $H(p,q)$ is a smooth real valued function with two variables p,q, where p stands for conjugate momentum and q is the generalized coordinate. Then the structure

$$\dot{q} = \frac{\partial H}{\partial p}, \quad \dot{p} = -\frac{\partial H}{\partial q}$$

is defined as the Hamiltonian system where H is the corresponding function, and $\dot{q}$ and $\dot{p}$ are Hamiltonian equations [3, 4, 5].

2.1.6.6 *Example*

Find the Hamiltonian function of the system $\ddot{x} = ax - bx^3$, where a, b are constants.

Solution: From Example 2.1.6.4, the dynamical system can be shown as

$$\begin{cases} \dot{x} = y, \\ \dot{y} = ax - bx^3. \end{cases} \tag{2.8}$$

To find the Hamiltonian function $H(x,y)$ of the system (2.8), one can consider

$$\dot{y} = -\frac{\partial H}{\partial x}, \quad \dot{x} = \frac{\partial H}{\partial y}.$$

This implies to give the Hamiltonian function as

$$H(x,y) = \frac{y^2}{2} - \left(\frac{ax^2}{2} - \frac{bx^4}{4} \right) = h(say).$$

The trajectories obtained in phase portrait of the system (2.8) develop to form different wave profiles of the system. The significant variation of the phase portraits of the system (2.8) is examined in (x,y) plane for different values of system parameters.

References

[1] L. Perko, Differential Equations and Dynamical Systems, Springer-Verlag, New York Inc. (2001).

[2] S. H. Strogatz, Nonlinear Dynamics and Chaos, Westview Press (USA) (2007).

[3] G. C. Layek, An Introduction to Dynamical Systems and Chaos, Springer, India (2015).

[4] S. N. Chow and J. K. Hale, Methods of Bifurcation Theory, Springer-Verlag, New York Berlin Heidelberg (1981).

[5] J. Guckenheimer and P. J. Holmes, Nonlinear Oscillations, Dynamical Systems and Bifurcations of Vector Fields, Springer-Verlag, New York (1983).

Chapter 3

Waves in Plasmas

3.1 Introduction to wave modes

Plasma is rich in wave phenomena. In any plasma system, plasma particles oscillate indiscriminately and interrelate among themselves in the presence of the electrostatic or electromagnetic forces and also acknowledge the externally exerted perturbations. For this fact, a collection of qualitatively different varieties of waves are available in plasmas [1]. The study on dynamical properties of different types of nonlinear waves in plasmas is an important research topic as such waves can be observed easily and also their theoretical background is well-established. Some of the important plasma waves include dust-acoustic (DA) wave, ion-acoustic (IA) wave, dust-ion-acoustic (DIA) wave, upper hybrid wave, electrostatic cyclotron wave and lower hybrid wave, etc.

3.1.1 Ion-acoustic (IA) waves

A sound wave can move from one layer to other layer due to collisions of the air molecules. If collisions are not there, then an ordinary sound wave can not originate. However, in the absence of collisions, one can discern such waves in plasmas. It is important to note that ions can transfer vibrations to each other due to their charges and, as a result, acoustic waves [2] can take place. As ions are massive in motion, they produce oscillations with low frequency. The corresponding dispersion relation for such ion-acoustic wave is as follows

$$\omega^2 \equiv k^2 c_s^2 \tag{3.1}$$

with the sound speed

$$c_s = \left[\frac{(\gamma_i T_{i0} + \gamma_e T_{e0})}{m_i}\right]^{1/2}, \tag{3.2}$$

where k, ω, T_{i0}, m_i and T_{e0} are, respectively, wave number, frequency of the ion-acoustic wave, ion temperature, mass of ions and electron temperature. Here, γ_j denotes the ratio between two specific heats ($j = i$ for ions and $j = e$ for electrons). The values of γ_e and γ_i are, in practice, influenced by temperature, region of density and wave number. If motions of ions are considered adiabatic in one dimension, then $\gamma_i = 3$, and if they are considered adiabatic compressions in three dimensions, then $\gamma_i = 5/3$. The term ion-acoustic comes due to the similarity observed in the dispersion relation of the IA wave and the corresponding relation for sound waves propagating through a gas.

3.1.2 Dust-acoustic (DA) waves

DA waves are very low frequency waves where the dust grains engage directly in the wave dynamics. DA waves were predicted theoretically by Rao et al. [3] in a multicomponent dusty plasma comprising of ions, electrons and negatively charged dusts. The phase velocity of DA waves is smaller than electron and ion thermal speeds. Hence, equilibrium in DA wave potential ϕ is established by inertialess ions and electrons. The inertia of dust particles plays an important role. The pressures of the electrons and ions provide the restoring force in DA waves, while the inertia is supplied by dust mass.

The corresponding dispersion relation is given by

$$\omega^2 = 3k^2 V_{Td}^2 + \frac{k^2 C_D^2}{1 + k^2 \lambda_D^2}, \tag{3.3}$$

where V_{Td} is dust thermal speed, k is wave number, and $C_D = \omega_{pd}\lambda_D$ is the DA speed with dusty plasma frequency ω_{pd} and Debye length λ_D. Since $\omega \gg kV_{Td}$, the DA wave frequency [3]

$$\omega = \frac{kC_D}{(1 + k^2 \lambda_D^2)^{1/2}}, \tag{3.4}$$

which in the long-wavelength limit (namely $k^2\lambda_D^2 << 1$) reduces to

$$\omega = kZ_{d0}\left(\frac{n_{d0}}{n_{i0}}\right)^{1/2}\left(\frac{k_B T_i}{m_d}\right)^{1/2}\left[1 + \frac{T_i}{T_e}\left(1 - \frac{Z_{d0}n_{d0}}{n_{i0}}\right)\right]^{-1/2}, \qquad (3.5)$$

which supports that the restoring force in DA waves arises from the pressures of the inertialess ions and electrons, while the inertia is supplied by dust mass to support the waves. Here $T_i, T_e, n_{i0}, n_{d0}, Z_{d0}, k_B$ and m_d are temperature of ions, temperature of electrons, unperturbed number density of ions, unperturbed number density of dust particles, unperturbed dust charge number, Boltzmann constant and mass of dust particles respectively. The frequency of the DA waves is much smaller as compared to the dust plasma frequency. The DA waves have been observed in several laboratory experiments [4], [5]. The observed DA wave frequencies are of the order of 10–20 Hz. Video images of the DA wavefronts are possible and they can be seen with the naked eye.

3.1.3 Dust-ion-acoustic (DIA) waves

The DIA waves were conceived by Shukla and Silin [6]. For DIA waves, the phase velocity is much larger (smaller) than the ion and dust thermal velocities (electron thermal velocity). In this case, the electron number density perturbation associated with the DIA waves is known, but the ion number density perturbation is determined from the ion continuity equation.

Considering DIA wave frequency $\omega >> kV_{Ti}, kV_{Td}$, the dispersion relation is given by

$$1 - \frac{k_{De}^2}{k^2} - \frac{\omega_{pi}^2 + \omega_{pd}^2}{\omega^2} = 0. \qquad (3.6)$$

As the dust grain has large mass, the ion plasma frequency ω_{pi} is much larger than the dust plasma frequency ω_{pd}. Here V_{Ti}, V_{Td} and k are ion thermal speed, dust thermal speed and wave number respectively. Thus, one can have

$$\omega^2 = \frac{k^2 C_S^2}{1 + k^2\lambda_{De}^2}, \qquad (3.7)$$

where $C_S = \omega_{pi}\lambda_{De}$. In the long wavelength limit, i.e., when

$$k^2\lambda_{De}^2 << 1, \qquad (3.8)$$

$$\text{then,} \quad \omega = k\omega_{pi}\lambda_{De}.$$

This shows that the phase velocity of the DIA waves in a dusty plasma is larger than $c_s = (k_B T_e/m_i)^{1/2}$. The signature of DIA waves is also found in laboratory experiments [7], [8]. The frequencies of the DIA waves for laboratory plasma are of the order kHz. At low frequencies, a new ultra low frequency mode arises from ion oscillations in static dust distribution.

3.1.4 Upper hybrid wave

If electrons oscillate perpendicular to the ambient magnetic field $\vec{B}_0$, then one can suppose that ions are so massive to oscillate at that frequency, so the ions form a fixed uniform background of positive charge. If one ignores the thermal motion of electrons, then $K_B T_e = 0$, where K_B is known as Boltzmann constant and T_e is temperature of electrons. In this case, at equilibrium, plasma has constant and uniform number density n_0 and ambient magnetic field $\vec{B}_0$, the initial velocity v_0 and the unperturbed electric field $\vec{E}_0 = 0$. If one considers only longitudinal wave, i.e., $\hat{k}$ is parallel to $\vec{E}_1$, then such waves are called upper hybrid waves and the corresponding frequency is called upper hybrid frequency.

If $\hat{k} \perp \vec{B}_0$, then the corresponding wave is called lower hybrid wave and the corresponding frequency is called lower hybrid frequency.

Equation of motion for the electrons is as follows

$$m\frac{d\vec{v}_e}{dt} = -e[\vec{E} + \vec{v}_e \times \vec{B}], \tag{3.9}$$

where $\vec{v}_e$ is velocity of electrons.

Equation of continuity gives

$$\frac{\partial n_e}{\partial t} + \vec{\nabla}.(n_e\vec{v}_e) = 0. \tag{3.10}$$

The Poisson equation gives

$$\varepsilon_0 \vec{\nabla}.\vec{E} = -e(n_e - n_i), \tag{3.11}$$

where m is mass of electrons, $\vec{E}$ denotes electric field, e is elementary charge, n_e is number density of electrons, n_i is number density of ions and $\vec{B}$ denotes magnetic field.

One can assume that there is no thermal motion, i.e., $K_B T_e = 0$, equilibrium plasma has constant number density n_0 and $E_0 = 0$ and $v_{e0} = 0$.

Now, one has to linearize the equations (3.9)-(3.11) by following relations:

$$n_e = n_0 + \varepsilon n_{e1},$$
$$\vec{B} = \vec{B}_0 + \varepsilon \vec{B}_1,$$
$$\vec{v} = 0 + \varepsilon \vec{v}_{e1}$$
$$\text{and} \quad \vec{E} = 0 + \varepsilon \vec{E}_1.$$

Then equation (3.9) becomes

$$m\varepsilon \frac{d\vec{v}_{e1}}{dt} = -e[\varepsilon \vec{E}_1 + \varepsilon \vec{v}_{e1} \times (\vec{B}_0 + \varepsilon \vec{B}_1)],$$

$$\text{or,} \quad m\varepsilon \frac{d\vec{v}_{e1}}{dt} = -e\varepsilon [\vec{E}_1 + \vec{v}_{e1} \times \vec{B}_0 + \varepsilon \vec{v}_{e1} \times \vec{B}_1].$$

Neglecting nonlinear term, one can get

$$m \frac{d\vec{v}_{e1}}{dt} = -e[\vec{E}_1 + \vec{v}_{e1} \times \vec{B}_0]. \qquad (3.12)$$

Now equation (3.10) becomes

$$\frac{\partial}{\partial t}(n_0 + \varepsilon n_{e1}) + \vec{\nabla}.\{(n_0 + \varepsilon n_{e1})\varepsilon \vec{v}_{e1}\} = 0,$$

$$\text{or,} \quad \frac{\partial n_0}{\partial t} + \varepsilon \frac{\partial n_{e1}}{\partial t} + \vec{\nabla}.(\varepsilon n_0 \vec{v}_{e1} + \varepsilon^2 n_{e1} \vec{v}_{e1}) = 0.$$

Neglecting higher order terms, the following relation can be obtained

$$\frac{\partial n_{e1}}{\partial t} + \vec{\nabla}.(n_0 \vec{v}_{e1}) = 0. \qquad (3.13)$$

Now Poisson equation (3.11) becomes

$$\varepsilon_0 \vec{\nabla}.(\varepsilon \vec{E}_1) = -e[n_0 + \varepsilon n_{e1} - n_0],$$

$$\text{or,} \quad \varepsilon_0 \vec{\nabla}.\vec{E}_1 = -e n_{e1}. \qquad (3.14)$$

One can contemplate only longitudinal waves, i.e., $\vec{K}$ is parallel to oscillatory electric field $\vec{E}_1$. Without any loss of generality, one can take $\vec{K}$ and $\vec{E}_1$ along x-direction, i.e.,

$$\vec{K} = (K_x, 0, 0) \text{ and } \vec{E}_1 = (E_x, 0, 0).$$

One can also assume that each quantity is proportional to $e^{i(kx-\omega t)}$. Now, one has to replace $\dfrac{\partial}{\partial x}$ by ik and $\dfrac{\partial}{\partial t}$ by $-i\omega$. So, surely B_0 is acting in z-direction. Therefore, one can consider $\vec{B}_0 = (0,0,B_0)$.

Then equation (3.12) becomes

$$m(-i\omega)\vec{v}_{e1} = -e[\vec{E}_1 + \vec{v}_{e1} \times \vec{B}_0],$$

or, $-mi\omega\vec{v}_{e1} = -e[(E_x,0,0) + (v_x,v_y,v_z) \times (0,0,B_0)].$

So, one can obtain following relations along the coordinate axes:

$$-mi\omega v_x = -e[E_x + v_y B_0]. \tag{3.15}$$

$$mi\omega v_y = -ev_x B_0. \tag{3.16}$$

$$-mi\omega v_z = 0 \implies v_z = 0. \tag{3.17}$$

From equation (3.18), one can obtain

$$v_y = -\frac{e}{mi\omega}v_x B_0. \tag{3.18}$$

Using equation (3.18) in equation (3.15), one can have

$$mi\omega v_x = e\left(E_x - \frac{e}{mi\omega}v_x B_0^2\right),$$

or, $\quad v_x\left(mi\omega + \dfrac{e^2 B_0^2}{mi\omega}\right) = eE_x,$

or, $\quad v_x = \dfrac{eE_x}{\left(mi\omega + \dfrac{e^2 B_0^2}{mi\omega}\right)},$

or, $\quad v_x = \dfrac{\dfrac{eE_x}{mi\omega}}{\left(1 - \dfrac{e^2 B_0^2}{m^2\omega^2}\right)},$

or, $\quad v_x = \dfrac{\dfrac{eE_x}{mi\omega}}{\left(1 - \dfrac{\omega_c^2}{\omega^2}\right)}, \tag{3.19}$

where $\omega_c = \dfrac{eB_0}{m}$ is the cyclotron frequency.

From equation (3.18), one formulates

$$v_y = -\frac{e}{mi\omega}B_0\left(\frac{\frac{eE_x}{mi\omega}}{1-\frac{\omega_c^2}{\omega^2}}\right),$$

or, $\quad v_y = -\left(\frac{e^2 B_0 E_x}{-m^2\omega^2}\right)\frac{1}{\left(1-\frac{\omega_c^2}{\omega^2}\right)},$

or, $\quad v_y = \left(\frac{\omega_c^2 E_x}{B_0\omega^2}\right)\frac{1}{\left(1-\frac{\omega_c^2}{\omega^2}\right)}.$ $\qquad$ (3.20)

From equation (3.13) (along *x*-direction), one gets

$$-i\omega n_{e1} + n_0(ik)v_x = 0,$$

or, $\qquad n_{e1} = \frac{k}{\omega}n_0 v_x.$ $\qquad$ (3.21)

From equation (3.14) (along *x*-direction), one can obtain

$$\varepsilon_0(ik)E_x = -en_{e1},$$

or, $\qquad \varepsilon_0(ik)E_x = -e\frac{k}{\omega}n_0 v_x,$ $\quad$ [using equation (3.21)]

or, $\quad \varepsilon_0(ik)E_x = -e\frac{k}{\omega}n_0\frac{\frac{eE_x}{mi\omega}}{\left(1-\frac{\omega_c^2}{\omega^2}\right)},$ $\quad$ [using equation (3.19)]

or, $\qquad E_x\left[\varepsilon_0 ik + \frac{e^2 k n_0}{mi\omega^2}\frac{1}{\left(1-\frac{\omega_c^2}{\omega^2}\right)}\right] = 0.$

Since $E_x \neq 0$, one can have

$$\varepsilon_0 ik + \frac{e^2 k n_0}{mi\omega^2}\frac{1}{\left(1-\frac{\omega_c^2}{\omega^2}\right)} = 0,$$

or, $\quad \left(1-\frac{\omega_c^2}{\omega^2}\right) = -\frac{e^2 k n_0}{mi\omega^2\varepsilon ik},$

or, $\quad \left(1-\frac{\omega_c^2}{\omega^2}\right) = \frac{e^2 n_0}{\varepsilon_0 m\omega^2},$

or, $\quad \left(1-\frac{\omega_c^2}{\omega^2}\right) = \frac{\omega_p^2}{\omega^2},$

where $\omega_p^2 = \dfrac{e^2 n_0}{\varepsilon_0 m} = \dfrac{4\pi e^2 n_0}{m}$,

$$\text{or,} \quad \omega^2 - \omega_c^2 = \omega_p^2,$$

$$\text{or,} \quad \omega = \sqrt{\omega_c^2 + \omega_p^2}, \tag{3.22}$$

which gives the frequency of the upper hybrid wave.

If $\omega_p = 0$, then $\omega = \omega_c$. Thus, if the plasma frequency ω_p is zero, then the upper hybrid frequency becomes cyclotron frequency.

3.1.5 Electrostatic cyclotron waves

Ion waves are perpendicular to $\vec{B}$. The equation of motion for ions is as follows

$$M\frac{dv_i}{dt} = e[\vec{E} + \vec{v}_i \times \vec{B}]. \tag{3.23}$$

The equation of continuity of ions gives

$$\frac{\partial n_i}{\partial t} + \frac{\partial}{\partial x}(n_i v_i) = 0, \tag{3.24}$$

where M is the mass of ions, $\vec{v}_i$ is the velocity of ions, $\vec{B}$ is the magnetic field and n_i is number density of ions.

To linearise equations (3.23) and (3.24), one has to split the dependent variables at two parts: equilibrium and perturbation parts. Thus, one can write

$$\begin{aligned}
\vec{E} &= -\vec{\nabla}\phi, \\
\vec{B} &= \vec{B}_0 + \vec{B}_1, \\
\phi &= \phi_0 + \phi_1 = 0 + \phi_1, \\
n_i &= n_0 + n_1 \\
\text{and} \quad \vec{v}_i &= \vec{v}_0 + \vec{v}_1 = 0 + \vec{v}_1.
\end{aligned} \tag{3.25}$$

Using equation (3.25) in equation (3.23), one can obtain

$$M\frac{d\vec{v}_i}{dt} = e[-\vec{\nabla}\phi_1 + \vec{v}_i \times (\vec{B}_0 + \vec{B}_1)],$$

$$\text{or,} \ M\frac{d\vec{v}_i}{dt} = e[-\vec{\nabla}\phi_1 + \vec{v}_i \times \vec{B}_0 + \vec{v}_i \times \vec{B}_1],$$

Removing nonlinear term $\vec{v}_1 \times \vec{B}_1$ in the above equation, the following relation is obtained

$$M\frac{d\vec{v}_1}{dt} = -e\vec{\nabla}\phi_1 + e\vec{v}_1 \times \vec{B}_0. \tag{3.26}$$

Now choosing $\vec{v}_1 = (v_x, v_y, v_z)$, $\phi_1 = \phi_x$, $\vec{B}_0 = (0,0,B_z)$, i.e., magnetic field is acting along z-direction, equation (3.26) becomes

$$M\left(\frac{dv_x}{dt}\hat{i} + \frac{dv_y}{dt}\hat{j} + \frac{dv_z}{dt}\hat{k}\right) = -e\left(\hat{i}\frac{\partial\phi_x}{\partial x}\right) + e(\hat{i}v_yB_z - \hat{j}v_xB_z).$$

Along the direction of coordinate axes, one can have the following relations:

$$M\frac{dv_x}{dt} = -e\frac{\partial\phi_x}{\partial x} + ev_yB_z. \tag{3.27}$$

$$M\frac{dv_y}{dt} = -ev_xB_z. \tag{3.28}$$

$$M\frac{dv_z}{dt} = 0. \tag{3.29}$$

Let all quantities be proportional to $e^{i(kx-\omega t)}$, where k denotes wave number and ω denotes frequency of the cyclotron wave.

Then $\dfrac{\partial}{\partial t} \equiv -i\omega$, $\dfrac{\partial}{\partial x} \equiv ik$ and $\dfrac{\partial^2}{\partial x^2} \equiv -k^2$.

From equation (3.27) one can get

$$-i\omega M v_x = -eik\phi_x + ev_yB_z. \tag{3.30}$$

From equation (3.28) one can have

$$-i\omega M v_y = -ev_xB_z,$$

$$\text{or, } v_y = \frac{ev_xB_z}{i\omega M}. \tag{3.31}$$

Using equation (3.31) in equation (3.30), one can obtain

$$-i\omega M v_x = -eik\phi_x + e\left(\frac{ev_x B_z}{i\omega M}\right)B_z,$$

$$\text{or,} \qquad -i\omega M v_x = -eik\phi_x + \left(\frac{e^2 B_z^2}{i\omega M}\right)v_x,$$

$$\text{or,} \qquad v_x\left(i\omega M + \frac{e^2 B_z^2}{i\omega M}\right) = eik\phi_x,$$

$$\text{or,} \qquad v_x = \frac{eik\phi_x}{i\omega M + \dfrac{e^2 B_z^2}{i\omega M}},$$

$$\text{or,} \qquad v_x = \frac{\dfrac{eik\phi_x}{i\omega M}}{1 - \dfrac{e^2 B_z^2}{\omega^2 M^2}},$$

$$\text{or,} \qquad v_x = \frac{\dfrac{ek\phi_x}{\omega M}}{\left(1 - \dfrac{\Omega_c^2}{\omega^2}\right)}, \qquad (3.32)$$

where $\Omega_c = \dfrac{eB_z}{m}$.

Using equation (3.32) in equation (3.31), one can have

$$v_y = \frac{eB_z}{i\omega M} \frac{\dfrac{ek\phi_x}{\omega M}}{\left(1 - \dfrac{\Omega_c^2}{\omega^2}\right)},$$

$$\text{or,} \qquad v_y = \frac{e^2 B_z k\phi_x}{i\omega^2 M^2} \frac{1}{\left(1 - \dfrac{\Omega_c^2}{\omega^2}\right)},$$

$$\text{or,} \qquad v_y = \frac{\Omega_c^2 k\phi_x}{i\omega^2 B_z} \frac{1}{\left(1 - \dfrac{\Omega_c^2}{\omega^2}\right)}, \qquad (3.33)$$

where $\Omega_c = \frac{eB_z}{M}$ is the ion cyclotron frequency.

Using equation (3.25) in equation (3.24), one can obtain

$$\frac{\partial}{\partial t}(n_0 + n_1) + \frac{\partial}{\partial x}((n_0 + n_1)(0 + v_1)) = 0,$$

or, $$\frac{\partial n_1}{\partial t} + \frac{\partial}{\partial x}(n_0 v_1) + \frac{\partial}{\partial x}(n_1 v_1) = 0.$$

Removing nonlinear term, one can get

$$\frac{\partial n_1}{\partial t} + \frac{\partial}{\partial x}(n_0 v_1) = 0. \tag{3.34}$$

Now all oscillating quantities can be expressed in the form $e^{i(kx - \omega t)}$, so $\frac{\partial}{\partial t} \equiv -i\omega$ and $\frac{\partial}{\partial x} \equiv ik$.

Then equation (3.34) becomes

$$-i\omega n_1 + n_0 i k v_x = 0,$$

or, $$n_1 = \frac{n_0 k v_x}{\omega}. \tag{3.35}$$

Let electrons follow Maxwell-Boltzmann distribution, then $n_e = n_0 exp\left(\frac{e\phi_x}{K_B T_e}\right)$, where K_B is the Boltzmann constant and T_e is temperature of electrons.

Initially

$$n_i = n_e,$$

or, $$n_0 + n_1 = n_0\left(1 + \frac{e\phi_x}{K_B T_e} + o(\phi_x^2) + ...\right),$$

or,

$$n_1 = \frac{n_0 e \phi_x}{K_B T_e} \text{ (Neglecting higher order terms),} \qquad (3.36)$$

or,

$$\frac{n_0 k v_x}{\omega} = \frac{n_0 e \phi_x}{K_B T_e}, \text{(by equation (3.35)),}$$

or,

$$v_x = \frac{e \phi_x \omega}{K_B T_e k},$$

or,

$$\frac{e k \phi_x}{\omega M} \cdot \frac{1}{\left(1 - \dfrac{\Omega_c^2}{\omega^2}\right)} = \frac{e \phi_x \omega}{K_B T_e k}, \text{(by equation (3.32)),}$$

or,

$$\left(1 - \frac{\Omega_c^2}{\omega^2}\right)^{-1} = \frac{\omega^2 M}{K_B T_e k^2},$$

or,

$$\left(1 - \frac{\Omega_c^2}{\omega^2}\right) = \frac{K_b T_e k^2}{\omega^2 M},$$

or,

$$1 - \frac{K_B T_e k^2}{\omega^2 M} = \frac{\Omega_c^2}{\omega^2},$$

or,

$$\frac{\omega^2 M - K_B T_e k^2}{M \omega^2} = \frac{\Omega_c^2}{\omega^2},$$

or,

$$\omega^2 - \frac{K_B T_e}{M} k^2 = \Omega_c^2,$$

or, $\quad \omega^2 - k^2 v_s^2 = \Omega_c^2, \left(\text{where } v_s = \sqrt{\dfrac{K_B T_e}{M}} \text{ is the ion-acoustic speed}\right)$

or,

$$\omega^2 = \Omega_c^2 + k^2 v_s^2, \qquad (3.37)$$

which is the required frequency of the electrostatic cyclotron wave.

If $v_s = 0$, then $\Omega_c^2 = \omega^2 \implies \Omega_c = \omega$, i.e., ion-cyclotron frequency = frequency of the cyclotron wave.

3.1.6 Lower hybrid wave

A lower hybrid wave is a longitudinal wave due to the oscillation of electrons and ions in a plasma with magnetic effect. The line of propagation must be very closely perpendicular to the stationary magnetic field (B_0). In this case, one can consider the motions of both electrons and ions.

If one considers the motion of ions, then

$$M\frac{d\vec{v}_i}{dt} = e[\vec{E} + \vec{v}_i \times \vec{B}].$$

(3.38)

The equation of continuity for ions is

$$\frac{\partial n_i}{\partial t} + \frac{\partial}{\partial x}(n_i v_i) = 0,$$

(3.39)

where M = mass of ions, $\vec{v}_i$ = velocity of ions, $\vec{B}$ = magnetic field, n_i = number density of ions, $\vec{E}$ = electric field, and e = elementary charge.

To linearise equations (3.38) and (3.39), one can write dependent variables as the sum of equilibrium and perturbed parts:

$$\begin{aligned}
\vec{B} &= \vec{B}_0 + \vec{B}_1, \\
\phi &= \phi_0 + \phi_1 = 0 + \phi_1, \\
n_i &= n_0 + n_1, \\
\vec{v}_i &= \vec{v}_0 + \vec{v}_1 = 0 + \vec{v}_1 \\
\text{and} \quad \vec{E} &= -\vec{\nabla}\phi.
\end{aligned}$$

(3.40)

Using equation (3.40) in equation (3.38), one can get

$$M\frac{d\vec{v}_1}{dt} = e[-\vec{\nabla}\phi + (0 + \vec{v}_1) \times (\vec{B}_0 + \vec{B}_1)],$$

or, $$M\frac{d\vec{v}_1}{dt} = e[-\vec{\nabla}\phi + \vec{v}_1 \times \vec{B}_0 + \vec{v}_1 \times \vec{B}_1].$$

(3.41)

Removing the nonlinear term, one can obtain

$$M\frac{d\vec{v}_1}{dt} = -e\vec{\nabla}\phi_1 + e\vec{v}_1 \times \vec{B}_0.$$

(3.42)

One can choose

$$\vec{v}_1 = (v_{ix}, v_{iy}, v_{iz}),$$
$$\text{and} \quad \vec{B}_0 = (0, 0, B_0),$$

i.e., magnetic field is acting along z-direction.

Now,

$$\vec{v}_1 \times \vec{B}_0 = \begin{vmatrix} \hat{i} & \hat{j} & \hat{k} \\ v_{ix} & v_{iy} & v_{iz} \\ 0 & 0 & B_0 \end{vmatrix}$$

$$= \hat{i}(v_{iy}B_0 - 0) - \hat{j}(v_{ix}B_0) + 0\hat{k}$$

Then, from equation (3.42), one can get

$$M\left(\frac{dv_{ix}}{dt}\hat{i} + \frac{dv_{iy}}{dt}\hat{j} + \frac{dv_{iz}}{dt}\hat{k}\right) = -e\left(\frac{\partial\phi_1}{\partial x}\hat{i} + 0\hat{j} + 0\hat{k}\right) + e\hat{i}v_{iy}B_0 - e\hat{j}v_{ix}B_0.$$

So, along the direction of coordinate axes, one can get

$$M\frac{dv_{ix}}{dt} = -e\frac{\partial\phi_1}{\partial x} + ev_{iy}B_0, \tag{3.43}$$

$$M\frac{dv_{iy}}{dt} = -ev_{ix}B_0, \tag{3.44}$$

$$M\frac{dv_{iz}}{dt} = 0. \tag{3.45}$$

Let all physical quantities vary proportional to $e^{i(kx-\omega t)}$, where k denotes wave number and ω denotes frequency of the lower hybrid wave.

Then

$$\frac{d}{dt} \equiv -i\omega, \quad \frac{\partial}{\partial x} \equiv ik \text{ and } \frac{\partial^2}{\partial x^2} \equiv -k^2. \tag{3.46}$$

Using equation (3.46) in equation (3.44)

$$-mi\omega v_{iy} = -ev_{ix}B_0,$$

$$\text{or, } v_{iy} = \frac{eB_0}{mi\omega}v_{ix}. \tag{3.47}$$

Using equation (3.46) in equation (3.43), one can get

$$-Mi\omega v_{ix} = -eik\phi_1 + ev_{iy}B_0,$$

or, $\qquad -Mi\omega v_{ix} = -eik\phi_1 + eB_0 \dfrac{eB_0}{Mi\omega} v_{ix},$ (by equation 3.47)

or, $\qquad v_{ix}\left[Mi\omega + \dfrac{e^2 B_0^2}{Mi\omega}\right] = eik\phi_1,$

or, $\qquad v_{ix}\left[1 + \dfrac{e^2 B_0^2}{(Mi\omega)^2}\right] = \dfrac{eik\phi_1}{Mi\omega},$

or, $\qquad v_{ix}\left[1 - \dfrac{e^2 B_0^2}{(M^2\omega^2)}\right] = \dfrac{ek\phi_1}{M\omega},$

or, $\quad v_{ix}\left[1 - \dfrac{\Omega_c^2}{(\omega^2)}\right] = \dfrac{ek\phi_1}{M\omega},$ where $\Omega_c = \dfrac{eB_0}{m}$ is the ion-cyclotron frequency,

or, $$v_{ix} = \dfrac{\dfrac{ek\phi_1}{M\omega}}{\left(1 - \dfrac{\Omega_c^2}{\omega^2}\right)} \qquad (3.48)$$

Now, using equation (3.40) in equation (3.39), one can obtain

$$\frac{\partial}{\partial t}(n_0 + n_i) + \frac{\partial}{\partial x}((n_0 + n_{i1})(0 + \vec{v}_1)) = 0,$$

or, $\qquad \dfrac{\partial n_{i1}}{\partial t} + \dfrac{\partial}{\partial x}(n_0 \vec{v}_1) = 0,$ (Neglecting $n_{i1}\vec{v}_1$),

or, $\qquad -\omega n_{i1} + n_0 ik\vec{v}_1 = 0,$

or, $\quad n_{i1} = \dfrac{n_0 k\vec{v}_1}{\omega}$ (Just for ions one can write $\vec{v}_1 = \vec{v}_{i1}$). $\qquad (3.49)$

If one considers the motions of electrons and does the similar approximation to obtain

$$v_{ex} = \frac{-\frac{ek}{m\omega}\phi_1}{\left(1 - \dfrac{\omega_c^2}{\omega^2}\right)}, \text{(where } \omega_c = \dfrac{eB_0}{m} \text{ is electron cyctron frequency)} \quad (3.50)$$

and $$n_{e1} = \frac{n_0 k v_{e1}}{\omega}. \qquad (3.51)$$

[For equations (3.50) and (3.51), one can see the part of upper hybrid wave.]

Now

$$n_i = n_e,$$

or,
$$n_{i1} = n_{e1},$$

or,
$$\frac{v_{i1} n_0 k}{\omega} = \frac{n_0 k v_{e1}}{\omega},$$

or,
$$\vec{v}_{i1} = \vec{v}_{e1},$$

or, $\quad v_{ix} = v_{ex}, \quad \text{(for x-components)}$

or,
$$\frac{\dfrac{ek}{M\omega}\phi_1}{\left(1 - \dfrac{\Omega_c^2}{\omega^2}\right)} = \frac{-\dfrac{ek}{m\omega}\phi_1}{\left(1 - \dfrac{\omega_c^2}{\omega^2}\right)}$$

or, $\quad m\left(1 - \dfrac{\omega_c^2}{\omega^2}\right) + M\left(1 - \dfrac{\Omega_c^2}{\omega^2}\right) = 0,$

or, $\quad m(\omega^2 - \omega_c^2) + M(\omega^2 - \Omega_c^2) = 0$

or, $\quad \omega^2(m + M) = \dfrac{M+m}{mM}e^2 B_0^2,$

or, $\quad \omega^2 = \dfrac{eB_0}{m}\dfrac{eB_0}{M},$

or, $\quad \omega^2 = \omega_c \Omega_c,$

or, $\quad \omega = \sqrt{\omega_c \Omega_c} \qquad\qquad (3.52)$

Thus, the lower hybrid frequency is the geometric mean between ion cyclotron frequency and electron cyclotron frequency.

3.2 Reductive perturbation technique and evolution equations

Reductive perturbation technique (RPT) is a method which can be used to derive nonlinear evolution equation from a set of model equations in plasmas for the study of nonlinear waves in small amplitude limit. Some of the standard nonlinear evolution equations in plasmas are KdV equation, KP equation, ZK equation, Burgers equation, Schamel equation, etc. In general, any plasma system consists of several dependent variables. In RPT, one can express nonlinear evolution equation in terms of a single dependent variable and its partial derivatives with respect to space variables and time. It rescales both time and space variables in the given system and introduces new space variables and time for describing the nonlinear wave phenomena with long wave length. It has a limitation that it relies on the experience for choosing specific scales [1]. In RPT, all dependent variables are expanded in terms of small perturbed parameter ε.

For example, one can consider expansions of number density (n) of plasma particles, its velocity (v), and corresponding electrostatic potential (ϕ) as

$$n = n_0 + n_1\varepsilon + n_2\varepsilon^2 + \cdots \tag{3.53}$$

$$v = 0 + v_1\varepsilon + v_2\varepsilon^2 + \cdots, \tag{3.54}$$

$$\phi = 0 + \phi_1\varepsilon + \phi_2\varepsilon^2 + \cdots. \tag{3.55}$$

Usually, these expansions are divided into two parts: unperturbed part and perturbed part. The unperturbed part is the first term of the expansion which can be acquired by the boundary conditions. In general, number density of plasma particles is perturbed about its equilibrium value n_0, which gives $n \to n_0$ as $x \to \pm\infty$. On the other hand, $v \to 0$ and $\phi \to 0$ as $x \to \pm\infty$. These conditions depend on physical situations. To obtain the dispersion relation for the plasma waves, it is required to consider the linearised version of the model equations. Any oscillatory physical quantity can be expressed as $e^{i\theta}$, with $\theta = kx - \omega(k)t$, where ω (frequency) and k (wave number) form a relationship which is known as the dispersion relation for the plasma wave. For propagation of long waves, one can consider $k = \varepsilon^p K$, where K denotes a new wave number of order one and p is a unrevealed constant to be obtained later. Then one can write

$$\theta(x,t) = \varepsilon^p Kx - \omega(\varepsilon^p K)\, t. \tag{3.56}$$

Since one is dealing with purely dispersive wave, the Taylor series expansion of $\omega(\varepsilon^p K)$ will contain only even or only odd powers K [1]. For nonlinear wave with long wave length only odd powers of K will materialize in the expansion. Then equation (3.56) can be expressed as

$$\theta(x,t) = \varepsilon^p K(x - \omega'(0)t) - \varepsilon^{3p} K^3 \omega'''(0)t + \cdots \tag{3.57}$$

Based on the relation (3.57), one can choose a suitable scaling for the independent variables x and t as

$$\xi = \varepsilon^p(x - \lambda t); \quad \tau = \varepsilon^{3p}t, \tag{3.58}$$

where λ is the phase velocity of the wave. Here ξ and τ are known as stretched variables. A suitable value of p is required to obtain a particular nonlinear evolution equation. For the KdV like equation, one needs to consider the value of p as $1/2$. One can suppose a linear dispersion law for an electrostatic mode as

$$\Sigma \frac{\omega_{pj}^2}{\omega^2 - k^2 v_{thj}^2} = 1. \tag{3.59}$$

For small values of ω and k, equation (3.59) can be approximated as

$$\omega = \lambda k - \frac{1}{A} k^3, \tag{3.60}$$

where ω_{pi} and ω_{pe} are ion plasma frequency and electron plasma frequency, v_{thi} and v_{the} are ion-thermal and electron-thermal speeds, respectively. Here, $A = 2\lambda \Sigma [\omega_{pj}^2 / (\lambda^2 - v_{thj}^2)^2]$ and $\frac{1}{A}$ will be coefficient of dispersion term in the KdV or mKdV equation. The phase for the nonlinear wave with long wave length is given by

$$kx - \omega t = k(x - \lambda t) + \frac{1}{A} k^3 t + \cdots, \tag{3.61}$$

and leads to the following standard stretching of the KdV equation as

$$\xi = \varepsilon^{1/2}(x - \lambda t); \ \tau = \varepsilon^{3/2} t. \tag{3.62}$$

Using the relation (3.62) with expansions (3.53)-(3.55), one can obtain the KdV equation for the considered plasma system.

3.2.1 The KdV equation

Consider an electron-ion plasma, where ions are mobile and electrons follow Maxwell distribution. The basic equations for the dynamics of ions are given by

$$\frac{\partial n_i}{\partial t} + \frac{\partial}{\partial x}(n_i u_i) = 0, \tag{3.63}$$

$$\frac{\partial u_i}{\partial t} + u_i \frac{\partial u_i}{\partial x} = -\frac{\partial \phi_i}{\partial x}, \tag{3.64}$$

$$\frac{\partial^2 \phi_i}{\partial x^2} = e^{\phi_i} - n_i, \tag{3.65}$$

where n_i is ion number density, u_i is ion velocity and ϕ_i is electrostatic wave potential. The equation (3.63) represents the equation of continuity of ions. The equation (3.64) represents the momentum balance. The third equation (3.65) is Poisson's equation. To derive the dispersion relation, one can linearize the system by considering the following expansions

$$
\begin{cases}
n_i = 1 + n, \\
u_i = 0 + u, \\
\phi_i = 0 + \phi.
\end{cases}
\tag{3.66}
$$

One can use the above expansion (3.66) in equation (3.63) and get

$$
\frac{\partial(1+n)}{\partial t} + \frac{\partial}{\partial x}[(1+n)u] = 0,
\tag{3.67}
$$

or, $\quad \dfrac{\partial n}{\partial t} + \dfrac{\partial u}{\partial x} = 0 \quad$ (neglecting nonlinear terms). $\tag{3.68}$

Similarly, using expansion (3.66) in equation (3.64), one can get

$$
\frac{\partial u}{\partial t} + u\frac{\partial u}{\partial x} = -\frac{\partial \phi}{\partial x},
\tag{3.69}
$$

or, $\quad \dfrac{\partial u}{\partial t} = -\dfrac{\partial \phi}{\partial x} \quad$ (neglecting nonlinear terms). $\tag{3.70}$

Finally, using expansion (3.66) in equation (3.65), one can obtain

$$
\frac{\partial^2 \phi}{\partial x^2} = 1 + \phi + \frac{\phi^2}{2!} + \cdots - (1+n)
\tag{3.71}
$$

or, $\quad \dfrac{\partial^2 \phi}{\partial x^2} = \phi - n, \quad$ (neglecting nonlinear terms). $\tag{3.72}$

To find the dispersion relation for low frequency wave, one can consider the perturbation in the form of $e^{i(kx-wt)}$, where $(kx-wt)$ is called the phase of the wave with k as wave number and w as its frequency.

Therefore, one considers the following perturbation expansions

$$
\begin{cases}
n = n_0 e^{i(kx-wt)}, \\
u = u_0 e^{i(kx-wt)}, \\
\phi = \phi_0 e^{i(kx-wt)}.
\end{cases}
\tag{3.73}
$$

Now, one can substitute equation (3.73) in equations (3.68), (3.70) and (3.72) to obtain the following relations:

$$-iwn + iku = 0. \tag{3.74}$$

$$-iwu + ik\phi = 0. \tag{3.75}$$

$$-k^2\phi = \phi - n. \tag{3.76}$$

Now, from equations (3.74) and (3.75), one can get

$$\begin{cases} n = \frac{ku}{w}, \\ \phi = \frac{uw}{k}. \end{cases} \tag{3.77}$$

From equations (3.76) and (3.77), one can obtain

$$-kw = \frac{w^2 - k^2}{kw}, \tag{3.78}$$

$$or, \ w^2 = \frac{k^2}{k^2 + 1}. \tag{3.79}$$

One can expand equation (3.79) to obtain w for small value of k as

$$w = k - \frac{1}{2}k^3 + \cdots. \tag{3.80}$$

Then, the phase of the wave becomes

$$kx - wt = k(x - t) + \frac{1}{2}k^3t. \tag{3.81}$$

Consider $k = \varepsilon^{\frac{1}{2}}$, where ε is a small parameter indicating the weakness of dispersion. Then, the stretched coordinates of space variable and time are given by:

$$\xi = \varepsilon^{\frac{1}{2}}(x - t), \quad \tau = \varepsilon^{\frac{3}{2}}t. \tag{3.82}$$

Expansions of dependent variables are as follows:

$$\begin{cases} n_i = 1 + \varepsilon n_1 + \varepsilon^2 n_2 + \cdots, \\ u_i = 0 + \varepsilon u_1 + \varepsilon^2 u_2 + \cdots, \\ \phi_i = 0 + \varepsilon \phi_1 + \varepsilon^2 \phi_2 + \cdots. \end{cases} \tag{3.83}$$

Now, using equation (3.82), one can get

$$\begin{cases} \frac{\partial}{\partial t} \equiv \varepsilon^{\frac{3}{2}}\frac{\partial}{\partial \tau} - \varepsilon^{\frac{1}{2}}\frac{\partial}{\partial \xi}, \\ \frac{\partial}{\partial x} \equiv \varepsilon^{\frac{1}{2}}\frac{\partial}{\partial \xi}. \end{cases} \tag{3.84}$$

One can substitute equations (3.83)-(3.84) in equation (3.63) to obtain

$$(\varepsilon^{\frac{3}{2}}\frac{\partial}{\partial \tau} - \varepsilon^{\frac{1}{2}}\frac{\partial}{\partial \xi})(1 + \varepsilon n_1 + \varepsilon^2 n_2 + \cdots) +$$

$$\varepsilon^{\frac{1}{2}}\frac{\partial}{\partial \xi}[(1 + \varepsilon n_1 + \varepsilon^2 n_2 + \cdots)(\varepsilon u_1 + \varepsilon^2 u_2 + \cdots)] = 0,$$

$$\Rightarrow \varepsilon^{\frac{3}{2}}(\frac{\partial u_1}{\partial \xi} - \frac{\partial n_1}{\partial \xi}) + \varepsilon^{\frac{5}{2}}[\frac{\partial n_1}{\partial \tau} - \frac{\partial n_2}{\partial \xi} + \frac{\partial u_2}{\partial \xi} + \frac{\partial}{\partial \xi}(u_1 n_1)] + \cdots = 0. \tag{3.85}$$

Using equations (3.83)-(3.84) in equation (3.64), one can get

$$(\varepsilon^{\frac{3}{2}}\frac{\partial}{\partial \tau} - \varepsilon^{\frac{1}{2}}\frac{\partial}{\partial \xi})(\varepsilon u_1 + \varepsilon^2 u_2 + \cdots) + \varepsilon^{\frac{1}{2}}(\varepsilon u_1 + \varepsilon^2 u_2 + \cdots)\frac{\partial}{\partial \xi}(\varepsilon u_1 + \varepsilon^2 u_2 + \cdots) =$$

$$-\varepsilon^{\frac{1}{2}}\frac{\partial}{\partial \xi}(\varepsilon \phi_1 + \varepsilon^2 \phi_2 + \cdots),$$

$$\Rightarrow \varepsilon^{\frac{3}{2}}[\frac{\partial \phi_1}{\partial \xi} - \frac{\partial u_1}{\partial \xi}] + \varepsilon^{\frac{5}{2}}(\frac{\partial u_1}{\partial \tau} - \frac{\partial u_2}{\partial \xi} + u_1\frac{\partial u_1}{\partial \xi} + \frac{\partial \phi_2}{\partial \xi}) + \cdots = 0. \tag{3.86}$$

Substituting equations (3.83)-(3.84) in equation (3.65), one can obtain

$$\varepsilon\frac{\partial^2}{\partial \xi^2}(\varepsilon \phi_1 + \varepsilon^2 \phi_2) = e^\phi - (1 + \varepsilon n_1 + \varepsilon^2 n_2),$$

$$\Rightarrow -\varepsilon(\phi_1 - n_1) + \varepsilon^2\left(\frac{\partial^2 \phi_1}{\partial \xi^2} - \phi_2 + n_2 - \frac{1}{2}\phi_1^2\right) + \varepsilon^3\frac{\partial^2 \phi_2}{\partial \xi^2} = 0. \tag{3.87}$$

Equating lowest order ε terms in equations (3.85)-(3.87), one can have

$$\begin{cases} \varepsilon^{\frac{3}{2}} : \frac{\partial u_1}{\partial \xi} - \frac{\partial n_1}{\partial \xi} = 0 \implies u_1 = n_1, \\ \varepsilon^{\frac{3}{2}} : \frac{\partial \phi_1}{\partial \xi} - \frac{\partial u_1}{\partial \xi} = 0 \implies \phi_1 = u_1, \\ \varepsilon : \phi_1 = n_1. \end{cases} \tag{3.88}$$

Equating the next higher order ε terms in equations (3.85)-(3.88), one gets

$$\varepsilon^{\frac{5}{2}} : \frac{\partial n_1}{\partial \tau} - \frac{\partial n_2}{\partial \xi} + \frac{\partial u_2}{\partial \xi} + \frac{\partial}{\partial \xi}(u_1 n_1) = 0, \tag{3.89}$$

$$\varepsilon^{\frac{5}{2}} : \frac{\partial u_1}{\partial \tau} - \frac{\partial u_2}{\partial \xi} + u_1 \frac{\partial u_1}{\partial \xi} + \frac{\partial \phi_2}{\partial \xi} = 0, \tag{3.90}$$

$$\varepsilon^2 : \frac{\partial^2 \phi_1}{\partial \xi^2} - \phi_2 + n_2 - \frac{1}{2}\phi_1^2 = 0. \tag{3.91}$$

Now, differentiating equation (3.91) w.r.t. ξ, one can get

$$\frac{\partial^3 \phi_1}{\partial \xi^3} - \frac{\partial \phi_2}{\partial \xi} + \frac{\partial n_2}{\partial \xi} - \phi_1 \frac{\partial \phi_1}{\partial \xi} = 0. \tag{3.92}$$

Using values of $\frac{\partial n_2}{\partial \xi}$ and $\frac{\partial \phi_2}{\partial \xi}$ from equations (3.89) and (3.90), respectively, in equation (3.92), one can obtain

$$\frac{\partial^3 \phi_1}{\partial \xi^3} - \frac{\partial u_2}{\partial \tau} - \frac{\partial u_1}{\partial \tau} - u_1 \frac{\partial u_1}{\partial \xi} + \frac{\partial n_1}{\partial \tau} - \frac{\partial n_2}{\partial \xi} + \frac{\partial}{\partial \xi}(u_1 n_1) - \phi_1 \frac{\partial \phi_1}{\partial \xi} = 0. \tag{3.93}$$

Then, one can eliminate 2nd order perturbed terms using equations (3.89) and (3.90). Also, using relation (3.88), one obtains the equation in only one variable ϕ_1 as

$$\frac{\partial^3 \phi_1}{\partial \xi^3} + 2\frac{\partial \phi_1}{\partial \tau} + 2\phi_1 \frac{\partial \phi_1}{\partial \xi} = 0. \tag{3.94}$$

The above equation can be rewritten as

$$\frac{\partial \phi_1}{\partial \tau} + \phi_1 \frac{\partial \phi_1}{\partial \xi} + \frac{1}{2}\frac{\partial^3 \phi_1}{\partial \xi^3} = 0. \tag{3.95}$$

Equation (3.95) is called the KdV equation for the Maxwellian electron-ion plasma.

3.2.2 The Burgers equation

The movement of dust-ion acoustic waves (DIAWs) is considered in nonextensive dusty plasmas composed of inertial singly charged ions, immobile negative dusts, Maxwellian positrons, and two electrons of lower and higher temperatures following q-nonextensive distribution. The basic equations for this plasma system are

$$\frac{\partial n}{\partial t} + \frac{\partial nu}{\partial x} = 0, \tag{3.96}$$

$$\frac{\partial u}{\partial t} + u\frac{\partial u}{\partial x} = -\frac{\partial \phi}{\partial x} + \eta\frac{\partial^2 u}{\partial x^2} \tag{3.97}$$

$$\frac{\partial^2 \phi}{\partial x^2} = -n + \mu + \mu_1 n_1 - \mu_p e^{-\sigma_2 \phi} + \mu_2 n_2 \tag{3.98}$$

where n, n_1 and n_2 are number densities of ions, lower and higher temperature electrons, respectively. The charge balance equation at equilibrium state is $n_0 + n_{p0} = n_{10} + n_{20} + Z_d n_{d0}$, where $n_{p0}(n_{d0})$ is the unperturbed positron (dust) number density, with "0" being the unperturbed quantities. Here, u is ion velocity, η is called the coefficient of viscosity and ϕ is electrostatic potential. The normalization of following quantities is given by: n_{pc}, n, n_1 and n_2 are normalized by n_{p0}, n_0, n_{10} and n_{20}, respectively. u is normalized by $C_i = (\frac{k_B T_e}{m_p})^{\frac{1}{2}}$ and ϕ is normalized by $\frac{k_B T_e}{e}$, with m_p as the mass of positrons, k_B is the Boltzmann constant, and e is electron charge. Here, $\mu_1 = \frac{n_{10}}{n_0}$, $\sigma_2 = \frac{T_1}{T_p}$, $\mu_2 = \frac{n_{20}}{n_0}$, $\mu = \frac{Z_d n_{d0}}{n_0} = 1 + \mu_p - \mu_1 - \mu_2$. It should be noted that T_p, T, T_1 and T_2 denote temperatures of positrons, ions, lower and higher temperature electrons, respectively. The time t is normalized by $\omega_{pi}^{-1} = (\frac{m_i}{4\pi n_0 e^2})^{\frac{1}{2}}$ and the space variable is normalized by the Debye length $\lambda_D = (\frac{k_B T_e}{4\pi n_0 e^2})^{\frac{1}{2}}$.

To describe nonextensive electron distribution, one can use the following function

$$f_e(v) = C_q \{1 + (q-1)[\frac{m_e v^2}{2k_B T_e} - \frac{e\phi}{k_B T_e}]\}^{\frac{1}{(q-1)}},$$

where, m_e and T_e are mass and temperature of electrons. Here, the function $f_e(v)$ extends the Tsallis entropy and satisfies the thermodynamics law. Thus, the normalization constant is specified by

$$C_q = n_{e0}\frac{\Gamma(\frac{1}{1-q})}{\Gamma(\frac{1}{1-q} - \frac{1}{2})}\sqrt{\frac{m_e(1-q)}{2\pi k_B T_e}} \text{ for } -1 < q < 1,$$

and

$$C_q = n_{e0}\frac{1+q}{2}\frac{\Gamma(\frac{1}{q-1} + \frac{1}{2})}{\Gamma(\frac{1}{q-1})}\sqrt{\frac{m_e(q-1)}{2\pi k_B T_e}} \text{ for } q > 1.$$

The following nonextensive electron number density is obtained after integrating $f_e(v)$ over all velocity spaces

$$n_e = n_{e0}\{1+(q-1)\tfrac{e\phi}{k_B T_e}\}^{1/(q-1)+1/2}.$$

Thus, the lower and higher temperature electrons in the nonextensive forms are taken as

$$n_1 = \{1+(q_1-1)\phi\}^{\frac{1}{q_1-1}+\frac{1}{2}} = \frac{1}{2}(q_1+1)\phi + \frac{1}{8}(q_1+1)(3-q_1)\phi^2, \qquad (3.99)$$

$$n_2 = \{1+(q_2-1)\sigma_1\phi\}^{\frac{1}{q_2-1}+\frac{1}{2}} = \frac{1}{2}\sigma_1(q_2+1)\phi + \sigma_1^2\frac{1}{8}(q_2+1)(3-q_2)\phi^2, \qquad (3.100)$$

where the parameter $q_i > -1$ $(i=1,2)$ quantifies the potential of nonextensivity and $\sigma_1 = \frac{T_1}{T_2}$.

The Burgers equation can be obtained by employing the reductive perturbation technique for which stretched coordinates for x and t are chosen as

$$\xi = \varepsilon(x-Vt), \quad \text{and} \quad \tau = \varepsilon^2 t, \qquad (3.101)$$

where the phase velocity of the DIA waves is given by V and ε is a smallness measure of dispersion. The expansions for n, u and ϕ are taken as

$$\begin{cases} n = 1+\varepsilon n_1+\varepsilon^2 n_2+\cdots \\ u = 0+\varepsilon u_1+\varepsilon^2 u_2+\cdots \\ \phi = 0+\varepsilon \phi_1+\varepsilon^2 \phi_2+\cdots. \end{cases} \qquad (3.102)$$

Now, substituting equations (3.101)-(3.102) into equation (3.96), one can compare the terms involving ε^2 and ε^3,

$$\varepsilon^2 : n_1 = \frac{1}{V}u_1, \qquad (3.103)$$

$$\varepsilon^3 : \frac{\partial n_1}{\partial \tau} - V\frac{\partial n_2}{\partial \xi} + \frac{\partial u_2}{\partial \xi} + \frac{\partial}{\partial \xi}(n_1 u_1) = 0. \qquad (3.104)$$

Similarly, using equations (3.101)-(3.102) into equation (3.97) and comparing the terms involving ε^2 and ε^3, one gets

$$\varepsilon^2 : u_1 = \frac{1}{V}\phi_1, \qquad (3.105)$$

$$\varepsilon^3 : \frac{\partial u_1}{\partial \tau} - V\frac{\partial u_2}{\partial \xi} + u_1\frac{\partial u_1}{\partial \xi} = -\frac{\partial \phi_2}{\partial \xi} + \eta\frac{\partial^2 u_1}{\partial \xi^2}. \qquad (3.106)$$

Finally, substituting equations (3.99)-(3.102) into equation (3.98), one can compare the terms involving ε and ε^2,

$$\varepsilon : \frac{1}{V^2} = a + \mu_p \sigma_2, \tag{3.107}$$

$$\varepsilon^2 : n_2 = \left(-\frac{\mu_p \sigma_2^2}{2} + b \right) \phi_1^2 + (\mu_p \sigma_2 + a) \phi_2, \tag{3.108}$$

where $a = \frac{1}{2}\{\mu_2 \sigma_1(q_2+1) + \mu_1(q_1+1)\}$ and $b = \frac{1}{8}\{\mu_2 \sigma_1^2(q_2+1)(3-q_2) + \mu_1(q_1+1)(3-q_1)\}$. Here, equation (3.107) is called the dispersion relation. Now, differentiating both sides of equation (3.108) w.r.t. ξ, one can get

$$\frac{\partial n_2}{\partial \xi} = \left(-\frac{\mu_p \sigma_2^2}{2} + b \right) \frac{\partial}{\partial \xi} \phi_1^2 + (\mu_p \sigma_2 + a) \frac{\partial}{\partial \xi} \phi_2. \tag{3.109}$$

Substituting equations (3.104) and (3.106) in equation (3.109), one can obtain

$$\frac{1}{V}\frac{\partial n_1}{\partial \tau} + \frac{1}{V}\frac{\partial}{\partial \xi}(n_1 u_1) + \frac{1}{V^2}\left(\frac{\partial u_1}{\partial \tau} + u_1 \frac{\partial u_1}{\partial \xi} + \frac{\partial \phi_2}{\partial \xi} - \eta \frac{\partial^2 u_1}{\partial \xi^2} \right) =$$
$$\left(-\frac{\mu_p \sigma_2^2}{2} + b \right) \frac{\partial}{\partial \xi} \phi_1^2 + (\mu_p \sigma_2 + a) \frac{\partial}{\partial \xi} \phi_2. \tag{3.110}$$

Using equations (3.103), (3.105) and (3.107) in equation (3.110) and eliminating all the second order perturbed terms, one simplifies the above equation as

$$\frac{\partial \phi_1}{\partial \tau} + \left(\frac{3}{2V} - (b - \mu_p \frac{\sigma_2^2}{2})V^3 \right) \phi_1 \frac{\partial \phi_1}{\partial \xi} = \frac{\eta}{2} \frac{\partial^2 \phi_1}{\partial \xi^2}. \tag{3.111}$$

Considering $\psi = \phi_1$, $M = \frac{3}{2V} - (b - \mu_p \frac{\sigma_2^2}{2})V^3$ and $N = \frac{\eta}{2}$, one can obtain the Burgers equation as

$$\frac{\partial \psi}{\partial \tau} + M\psi \frac{\partial \psi}{\partial \xi} = N \frac{\partial^2 \psi}{\partial \xi^2}. \tag{3.112}$$

3.2.3 The KP equation

Consider a magnetized plasma constituting cold ions and nonthermal electrons featuring Cairns-Tsallis distribution for ion-acoustic waves (IAWs), where an external static magnetic field $\mathbf{B} = \hat{x}B_0$ is acting along the direction of the x-axis. Here, $\hat{x}$ denotes a unit vector in the direction x-axis and B_0

is unperturbed strength of magnetic field. The normalized basic equations for IAWs propagating in the xy-plane are

$$\frac{\partial n}{\partial t} + \nabla \cdot (n\tilde{U}) = 0, \tag{3.113}$$

$$\frac{\partial \tilde{U}}{\partial t} + (\tilde{U} \cdot \nabla)\tilde{U} = -\nabla \phi + \tilde{U} \times \hat{x}, \tag{3.114}$$

$$\nabla^2 \phi = \alpha_1 (n_e - n). \tag{3.115}$$

The normalized form electron density is given by

$$n_e = (1 + K_1 \phi + K_2 \phi^2)\{1 + (q-1)\phi\}^{\frac{1}{q-1}+\frac{1}{2}}, \tag{3.116}$$

where $K_1 = -\frac{16q\alpha_1}{(5q-3)(3q-1)+12\alpha_1}$ and $K_2 = \frac{16q(2q-1)\alpha_1}{(5q-3)(3q-1)+12\alpha_1}$.

Here, n and n_e are the ion and electron densities, respectively. $\tilde{U} = (u, v, w)$ is ion velocity and ϕ stands for electrostatic potential. The parameter $\alpha_1 = \frac{r^2}{\lambda^2}$, where $r = \frac{C_s}{\Omega}$ and $\Omega = \frac{eB_0}{mc}$ are ions' gyro-radius, and gyrofrequency, respectively. Here, c represents the speed of the light, m denotes mass of ions and e denotes the electronic charge. $C_s = (T_e/m)^{1/2}$ is IA velocity and $\lambda = \sqrt{T_e/4\pi e^2 n_0}$ denotes electron Debye length, where T_e signifies the temperature of electrons. n_0, and n_{e0} are number densities of ions and electrons at unperturbed state, respectively.

The normalization used for quantities in equations (3.113)-(3.115) are as follows: $\tilde{U}^\star = \frac{\tilde{U}}{C_s}$, $\phi^\star = \frac{e\phi}{T_e}$, $x^\star = \frac{x}{r}$, $y^\star = \frac{y}{r}$ and $t^\star = \frac{t}{\Omega^{-1}}$. Here, $\star$ is removed in equations (3.113)-(3.115) for simplicity.

One can write equations (3.113)-(3.115) in component form as:

$$\frac{\partial n}{\partial t} + \frac{\partial (nu)}{\partial x} + \frac{\partial (nv)}{\partial y} = 0, \tag{3.117}$$

$$\frac{\partial u}{\partial t} + (u\frac{\partial}{\partial x} + v\frac{\partial}{\partial y})u = -\frac{\partial \phi}{\partial x}, \tag{3.118}$$

$$\frac{\partial v}{\partial t} + (u\frac{\partial}{\partial x} + v\frac{\partial}{\partial y})v = -\frac{\partial \phi}{\partial y} + w, \tag{3.119}$$

$$\frac{\partial w}{\partial t} + (u\frac{\partial}{\partial x} + v\frac{\partial}{\partial y})w = -v, \tag{3.120}$$

$$\left(\frac{\partial^2}{\partial x^2} + \frac{\partial^2}{\partial y^2}\right)\phi = \alpha_1[(1 + K_1\phi + K_2\phi^2)\{1 + (q-1)\}^{\frac{1}{q-1}+\frac{1}{2}} - n]. \tag{3.121}$$

One can apply the concept of RPT to obtain the KP equation. In agreement with RPT, the new stretching for space and time variables are:

$$\begin{cases} Y = \varepsilon^2 y, \\ \eta = \varepsilon(x - Vt), \\ \tau = \varepsilon^3 t, \end{cases} \tag{3.122}$$

where, V represents phase velocity of IAW, and the parameter ε characterizes the extent of the nonlinearity, such that $0 < \varepsilon \ll 1$. One considers the following expansions for the dependent variables involved in equations (3.117)-(3.121)

$$\begin{cases} n = 1 + \varepsilon^2 n_1 + \varepsilon^4 n_2 + \cdots \\ u = \varepsilon^2 u_1 + \varepsilon^4 u_2 + \cdots \\ v = \varepsilon^3 v_1 + \varepsilon^5 v_2 + \cdots \\ w = \varepsilon^3 w_1 + \varepsilon^5 w_2 + \cdots \\ \phi = \varepsilon^2 \phi_1 + \varepsilon^4 \phi_2 + \cdots \end{cases} \tag{3.123}$$

Substituting equations (3.122)-(3.123) into the normalized equation (3.117) and comparing the coefficient of ε^3 and ε^5, one can get

$$n_1 = \frac{1}{V} u_1, \tag{3.124}$$

$$\frac{\partial n_1}{\partial \tau} - V \frac{\partial n_2}{\partial \eta} + \frac{\partial v_1}{\partial Y} + \frac{\partial}{\partial \eta}(n_1 u_1) + \frac{\partial u_2}{\partial \eta} = 0. \tag{3.125}$$

Using equations (3.122)-(3.123) in equation (3.118) and comparing the coefficient of ε^3 and ε^5, one can obtain

$$u_1 = \frac{1}{V} \phi_1, \tag{3.126}$$

$$\frac{\partial u_1}{\partial \tau} - V \frac{\partial u_2}{\partial \eta} + u_1 \frac{\partial u_1}{\partial \eta} = -\frac{\partial \phi_2}{\partial \eta}. \tag{3.127}$$

From equations (3.122), (3.123) and (3.119), one can have the following relation by comparing the coefficients of different orders of ε

$$V \frac{\partial v_1}{\partial \eta} = \frac{\partial \phi_1}{\partial Y}, \tag{3.128}$$

$$\frac{\partial v_1}{\partial \tau} - V \frac{\partial v_2}{\partial \eta} + u_1 \frac{\partial v_1}{\partial \eta} = -\frac{\partial \phi_2}{\partial Y}. \tag{3.129}$$

Using equations (3.122), (3.123) and (3.120) and comparing the coefficient of ε^5, one can obtain

$$\frac{\partial w_1}{\partial \tau} - V \frac{\partial w_2}{\partial \eta} + u_1 \frac{\partial w_1}{\partial \eta} = 0. \tag{3.130}$$

Using equations (3.122), (3.123) and (3.121) and on comparing the coefficients of orders of ε^2 and ε^3, one can get

$$\frac{\phi_1}{n_1} = \frac{1}{a + K_1}, \tag{3.131}$$

$$\frac{\partial^2 \phi_1}{\partial \eta^2} = \alpha_1 (P\phi_2 + Q\phi_1^2 - n_2), \tag{3.132}$$

where $P = a + K_1$, $Q = b + K_2 + aK_1$, $a = \frac{q+1}{2}$ and $b = \frac{(q+1)(3-q)}{8}$.

Differentiating relation (3.132) with respect to η, one gets

$$\frac{\partial^3 \phi_1}{\partial \eta^3} = \alpha_1 \left(P \frac{\partial \phi_2}{\partial \eta} + 2Q\phi_1 \frac{\partial \phi_1}{\partial \eta} - \frac{\partial n_2}{\partial \eta} \right), \tag{3.133}$$

or, $\quad \frac{1}{\alpha_1} \frac{\partial^3 \phi_1}{\partial \eta^3} = P \left(V \frac{\partial u_2}{\partial \eta} - \frac{\partial u_1}{\partial \tau} - u_1 \frac{\partial u_1}{\partial \eta} \right) + 2Q\phi_1 \frac{\partial \phi_1}{\partial \eta} - \frac{\partial n_2}{\partial \eta}. \quad$ (3.134)

$$\text{(Using equation (3.127))}$$

From equations (3.124) and (3.126)

$$V^2 = \frac{\phi_1}{n_1},$$

or, $\quad V^2 = \frac{1}{a + K_1}, \qquad \text{(Using equation (3.131))} \quad (3.135)$

or, $\quad V^2 = \frac{1}{P}. \tag{3.136}$

From equations (3.134), (3.136) and (3.127), one can get

$$\frac{1}{\alpha_1} \frac{\partial^3 \phi_1}{\partial \eta^3} = \frac{1}{V} \frac{\partial u_2}{\partial \eta} - \frac{\partial n_2}{\partial \eta} - P \frac{\partial u_1}{\partial \tau} - P \left(\frac{1}{V} \phi_1 \right) \frac{\partial}{\partial \eta} \left(\frac{1}{V} \phi_1 \right) + 2Q\phi_1 \frac{\partial \phi_1}{\partial \eta},$$

or, $\quad \frac{1}{\alpha_1} \frac{\partial^3 \phi_1}{\partial \eta^3} = -P \frac{\partial u_1}{\partial \tau} + \frac{1}{V} \left(\frac{\partial u_2}{\partial \eta} - V \frac{\partial n_2}{\partial \eta} \right) - \frac{P}{V^2} \phi_1 \frac{\partial \phi_1}{\partial \eta} + 2Q\phi_1 \frac{\partial \phi_1}{\partial \eta},$

or, $\quad \frac{1}{\alpha_1} \frac{\partial^3 \phi_1}{\partial \eta^3} + P \frac{\partial u_1}{\partial \tau} = \frac{1}{V} \left(-\frac{\partial n_1}{\partial \tau} - \frac{\partial v_1}{\partial Y} + \frac{\partial}{\partial \eta} (n_1 u_1) \right) - \frac{P}{V^2} \phi_1 \frac{\partial \phi_1}{\partial \eta} + 2Q\phi_1 \frac{\partial \phi_1}{\partial \eta},$

$$\text{(Using equation (3.125))}$$

or, $\dfrac{1}{\alpha_1}\dfrac{\partial^3 \phi_1}{\partial \eta^3}+P\dfrac{\partial u_1}{\partial \tau}=-\dfrac{1}{V}\dfrac{\partial}{\partial \tau}\left(\dfrac{1}{V^2}\phi_1\right)-\dfrac{1}{V}\dfrac{\partial v_1}{\partial Y}-\dfrac{1}{V}\dfrac{\partial}{\partial \eta}\left(\dfrac{1}{V^3}\phi_1^2\right)+\phi_1\dfrac{\partial \phi_1}{\partial \eta}\left(2Q-\dfrac{P}{V^2}\right),$

(Using equations (3.124) and (3.126))

or, $\dfrac{1}{\alpha_1}\dfrac{\partial^3 \phi_1}{\partial \eta^3}+P\dfrac{\partial u_1}{\partial \tau}=-\dfrac{1}{V^3}\dfrac{\partial \phi_1}{\partial \tau}-\dfrac{1}{V}\dfrac{\partial v_1}{\partial Y}-\dfrac{2}{V^4}\phi_1\dfrac{\partial \phi_1}{\partial \eta}+\phi_1\dfrac{\partial \phi_1}{\partial \eta}\left(2Q-\dfrac{P}{V^2}\right),$

or, $\dfrac{1}{\alpha_1}\dfrac{\partial^3 \phi_1}{\partial \eta^3}+P\dfrac{\partial}{\partial \tau}\left(\dfrac{1}{V}\phi_1\right)=-\dfrac{P}{V}\dfrac{\partial \phi_1}{\partial \tau}-\dfrac{1}{V}\dfrac{\partial v_1}{\partial Y}+\phi_1\dfrac{\partial \phi_1}{\partial \eta}\left(2Q-\dfrac{P}{V^2}-\dfrac{2}{V^4}\right),$

(Using equation (3.136))

or, $\dfrac{1}{\alpha_1}\dfrac{\partial^3 \phi_1}{\partial \eta^3}+\dfrac{2P}{V}\dfrac{\partial \phi_1}{\partial \tau}=\phi_1\dfrac{\partial \phi_1}{\partial \eta}\left(2Q-P^2-2P^2\right)-\dfrac{1}{V}\dfrac{\partial v_1}{\partial Y},$ (Using equation (3.136))

or, $\dfrac{1}{\alpha_1}\dfrac{\partial^3 \phi_1}{\partial \eta^3}-(2Q-3P^2)\phi_1\dfrac{\partial \phi_1}{\partial \eta}+\dfrac{2P}{V}\dfrac{\partial \phi_1}{\partial \tau}=-\dfrac{1}{V}\dfrac{\partial v_1}{\partial Y},$

or, $\dfrac{\partial \phi_1}{\partial \tau}-\dfrac{V}{2P}(2Q-3P^2)\phi_1\dfrac{\partial \phi_1}{\partial \eta}+\dfrac{V}{2P\alpha_1}\dfrac{\partial^3 \phi_1}{\partial \eta^3}=-\dfrac{V^2}{2}\dfrac{\partial v_1}{\partial Y},$ (3.137)

Differentiating above equation (3.137) with respect to η, one can obtain

$$\dfrac{\partial}{\partial \eta}\left(\dfrac{\partial \phi_1}{\partial \tau}-\dfrac{V}{2P}(2Q-3P^2)\phi_1\dfrac{\partial \phi_1}{\partial \eta}+\dfrac{V}{2P\alpha_1}\dfrac{\partial^3 \phi_1}{\partial \eta^3}\right)=-\dfrac{V^2}{2}\dfrac{\partial^2 v_1}{\partial \eta \partial Y},$$

or, $\dfrac{\partial}{\partial \eta}\left(\dfrac{\partial \phi_1}{\partial \tau}-\dfrac{V}{2P}(2Q-3P^2)\phi_1\dfrac{\partial \phi_1}{\partial \eta}+\dfrac{V}{2P\alpha_1}\dfrac{\partial^3 \phi_1}{\partial \eta^3}\right)=-\dfrac{V^2}{2}\dfrac{\partial}{\partial Y}\left(\dfrac{1}{V}\dfrac{\partial \phi_1}{\partial Y}\right),$

(Using equation (3.128))

or, $\dfrac{\partial}{\partial \eta}\left(\dfrac{\partial \phi_1}{\partial \tau}-\dfrac{V}{2P}(2Q-3P^2)\phi_1\dfrac{\partial \phi_1}{\partial \eta}+\dfrac{V}{2P\alpha_1}\dfrac{\partial^3 \phi_1}{\partial \eta^3}\right)=-\dfrac{V}{2}\dfrac{\partial^2 \phi_1}{\partial Y^2},$

or, $\dfrac{\partial}{\partial \eta}\left(\dfrac{\partial \phi_1}{\partial \tau}-A\phi_1\dfrac{\partial \phi_1}{\partial \eta}+B\dfrac{\partial^3 \phi_1}{\partial \eta^3}\right)+C\dfrac{\partial^2 \phi_1}{\partial Y^2}=0.$ (3.138)

Equation (3.138) is the KP equation for considered plasma system, where $A=\dfrac{V}{2P}(3P^2-2Q)$, $B=\dfrac{V}{2\alpha_1 P}$ and $C=\dfrac{V}{2}$.

3.2.4 The ZK and mZK equations

Consider a plasma system that is composed of negative cold dust particles with q-nonextensive distributed hot ions and electrons in presence of stationary neutrals. Furthermore, an external static magnetic field $\mathbf{B_0}=B_0\hat{x}$ is considered acting in the direction of x-axis. The dynamics of dust-acoustic waves are described by equations

$$\frac{\partial n_d}{\partial t} + \nabla n_d u_d = 0, \tag{3.139}$$

$$\frac{\partial u_d}{\partial t} + u_d \cdot \nabla u_d = \nabla \phi - u_d \times \Omega \hat{x}, \tag{3.140}$$

$$\nabla^2 \phi = n_d - \mu_i n_i + \mu_e n_e, \tag{3.141}$$

where n_d, n_e, n_i are number densities of dusts, electrons, ions and u_d is velocity of dusts, Ω is dust cyclotron frequency and ϕ is electrostatic potential. One can have $\delta_i = \frac{T_{eff}}{T_i Z_d}$ and $\delta_e = \frac{T_{eff}}{T_e Z_d}$ with $T_{i(e)}$ as temperatures of ions (electrons) and $T_{eff} = \frac{Z_d T_i T_e}{(\mu_i T_e + \mu_e T_i)}$ is effective temperature. Here, Z_d is the charge number. Also, $\mu_i = \frac{n_{i0}}{n_{d0} Z_d}$ and $\mu_e = \frac{n_{e0}}{n_{d0} Z_d}$. The normalization factors used are: n_{d0} and $Z_d n_{d0}$ for n_d and $n_{i,e}$, respectively, $\left(\frac{T_{eff}}{m_d}\right)^{\frac{1}{2}}$ for u_d, $\frac{T_{eff}}{e Z_d}$ for ϕ, $\left(\frac{m_d}{4\pi n_{d0} e^2 Z_d^2}\right)^{\frac{1}{2}}$ for t, $\left(\frac{4\pi n_{d0} e^2 Z_d^2}{T_{eff}}\right)^{-\frac{1}{2}}$ for ∇ and $\left(\frac{m_d}{4\pi n_{d0} e^2 Z_d^2}\right)^{\frac{1}{2}}$ for Ω.

The nonextensive distributive forms of ions and electrons are obtained in detail in the work [9]. Thus, dimensionless electron and ion number densities as:

$$\begin{cases} n_i = \{1 - \delta_i(q_i - 1)\phi\}^{\frac{1}{q_i-1}+\frac{1}{2}}, \\ n_e = \{1 + \delta_e(q_e - 1)\phi\}^{\frac{1}{q_e-1}+\frac{1}{2}}, \end{cases} \tag{3.142}$$

with q_i and q_e being the nonextensive parameters of ions and electrons. Here, $q_e, q_i > -1$ measure the potential of nonextensivity.

One can use the reductive perturbation method to examine DAWs based on the ZK equation using the following stretched variables

$$X = \varepsilon^{\frac{1}{2}}(x - \lambda t), \quad Y = \varepsilon^{\frac{1}{2}} y, \quad Z = \varepsilon^{\frac{1}{2}} z, \quad \tau = \varepsilon^{\frac{3}{2}} t. \tag{3.143}$$

Here, ε is measure of dispersion weakness and λ is the DAW phase velocity. The expansions of other quantities are given as

$$\begin{cases} n_d = 1 + \varepsilon n_{d1} + \varepsilon^2 n_{d2} + \varepsilon^3 n_{d3} + \cdots, \\ u_{dx} = 0 + \varepsilon u_{dx1} + \varepsilon^2 u_{dx2} + \varepsilon^3 u_{dx3} + \cdots, \\ u_{dy} = 0 + \varepsilon^{\frac{3}{2}} u_{dy1} + \varepsilon^2 u_{dy2} + \varepsilon^{\frac{5}{2}} u_{dy3} + \cdots, \\ u_{dz} = 0 + \varepsilon^{\frac{3}{2}} u_{dz1} + \varepsilon^2 u_{dz2} + \varepsilon^{\frac{5}{2}} u_{dz3} + \cdots, \\ \phi = 0 + \varepsilon \phi_1 + \varepsilon^2 \phi_2 + \varepsilon^3 \phi_3 + \cdots. \end{cases} \tag{3.144}$$

Substituting equations (3.142)-(3.144) into equation (3.139), one can get

$$\varepsilon^{\frac{3}{2}} : n_{d1} = \frac{1}{\lambda} u_{dx1}, \quad (3.145)$$

$$\varepsilon^{\frac{5}{2}} : -\lambda \frac{\partial n_{d2}}{\partial X} + \frac{\partial n_{d1}}{\partial T} + \frac{\partial}{\partial X}(u_{dx2} + n_{dx1}u_{dx1}) + \frac{\partial u_{dy2}}{\partial Y} + \frac{\partial u_{dz2}}{\partial Z} = 0. \quad (3.146)$$

Substituting equations (3.142)-(3.144) into equation (3.140), one can get

$$\varepsilon^{\frac{3}{2}} : u_{dx1} = -\frac{1}{\lambda} \phi_1, \quad (3.147)$$

$$\varepsilon^{\frac{5}{2}} : -\lambda \frac{\partial}{\partial X} u_{dx2} + \frac{\partial}{\partial T} u_{dx1} + u_{dx1} \frac{\partial}{\partial X} u_{dx1} - \frac{\partial}{\partial X} \phi_2 = 0, \quad (3.148)$$

$$u_{dy1} = -\frac{1}{\Omega} \frac{\partial}{\partial Z} \phi_1, \quad u_{dy2} = -\frac{\lambda}{\Omega^2} \frac{\partial^2}{\partial X \partial Y} \phi_1,$$

$$u_{dz1} = \frac{1}{\Omega} \frac{\partial}{\partial Y} \phi_1, \quad u_{dz2} = -\frac{\lambda}{\Omega^2} \frac{\partial^2}{\partial X \partial Z} \phi_1. \quad (3.149)$$

Substituting equations (3.142)-(3.144) into equation (3.141), one can get

$$\varepsilon^0 : \mu_i = \mu_e + 1, \quad (3.150)$$

$$\varepsilon : \lambda = \sqrt{\frac{2(\mu_e T_i + \mu_i T_e)}{\mu_i (q_i + 1) T_e + \mu_e (q_e + 1) T_i}}, \quad (3.151)$$

$$\varepsilon^2 : \left(\frac{\partial^2}{\partial X^2} + \frac{\partial^2}{\partial Y^2} + \frac{\partial^2}{\partial Z^2} \right) \phi_1 - n_{d2} - A\phi_2 + B\phi_1^2 = 0, \quad (3.152)$$

where $A = \frac{1}{2}[\mu_i(q_i + 1)\delta_i + \mu_e(q_e + 1)\delta_e]$, and
$B = \frac{1}{8}[\mu_e(q_e + 1)(3 - q_e)\delta_e^2 - \mu_i(q_i + 1)(3 - q_i)\delta_i^2]$.

As q_i and q_e converge to 1, one can obtain $\lambda = 1$. In that case, this study holds a good agreement with the work [10].

Now, differentiating both sides of equation (3.152) w.r.t. X, one can get

$$\frac{\partial}{\partial X} \left(\frac{\partial^2}{\partial X^2} + \frac{\partial^2}{\partial Y^2} + \frac{\partial^2}{\partial Z^2} \right) \phi_1 - \frac{\partial}{\partial X} n_{d2} - A \frac{\partial}{\partial X} \phi_2 + B \frac{\partial}{\partial X} \phi_1^2 = 0. \quad (3.153)$$

One can replace all second order perturbation terms using equations (3.146) and (3.148) and replace first order perturbed terms using equations (3.147)-(3.149). Then, one can obtain the following equation as the ZK equation (ZK) in one variable $\psi = \phi_1$

$$\frac{\partial \psi}{\partial \tau} + C\psi \frac{\partial \psi}{\partial X} + \frac{\lambda^3}{2} \frac{\partial^3 \psi}{\partial X^3} + \frac{\lambda^3}{2}\left(1 + \frac{1}{\Omega^2 \lambda}\right)\left(\frac{\partial^3 \psi}{\partial X \partial Y^2} + \frac{\partial^3 \psi}{\partial X \partial Z^2}\right) = 0,$$

(3.154)

where $C = B\lambda^3 - \frac{3}{2\lambda}$.

The modified ZK equation is derived in a situation when the nonlinear coefficient C in the ZK equation is approximately zero for certain values of parameters. In such case, the ZK equation is not valid to study nonlinear waves. Therefore, one can obtain the modified ZK equation utilizing new stretched variables

$$X = \varepsilon(x - \lambda t), \quad Y = \varepsilon y, \quad Z = \varepsilon z, \quad \tau = \varepsilon^3 t. \quad (3.155)$$

The expansions of n_d, u_{dx} and ϕ follow from equation (3.144), while u_{dy}, u_{dz} are given as

$$\begin{cases} u_{dy} = \varepsilon^2 u_{dy1} + \varepsilon^3 u_{dy2} + \varepsilon^4 u_{dy3} + \cdots, \\ u_{dz} = \varepsilon^2 u_{dz1} + \varepsilon^3 u_{dz2} + \varepsilon^4 u_{dz3} + \cdots. \end{cases} \quad (3.156)$$

Substituting equations (3.155)-(3.156) into equation (3.139), one can get

$$\varepsilon^4 : -\lambda \frac{\partial}{\partial X} n_{d3} + \frac{\partial}{\partial T} n_{d1} + \frac{\partial}{\partial X}(u_{dx3} + n_{d1}u_{dx2} + n_{dx2}u_{dx1}) + \quad (3.157)$$

$$\frac{\partial}{\partial Y}(u_{dy2} + n_{d1}u_{dy1}) + \frac{\partial}{\partial Z}(u_{dz2} + n_{d1}u_{dz1}) = 0.$$

Substituting equations (3.155)-(3.156) into equation (3.140), one can get

$$\varepsilon^3 : u_{dx2} = -\frac{1}{2\lambda}u_{dx1}^2 - \frac{1}{\lambda}\phi_2, \quad (3.158)$$

$$\varepsilon^4 : -\lambda \frac{\partial}{\partial X}u_{dx3} + \frac{\partial}{\partial T}u_{dx1} + u_{dx1}\frac{\partial}{\partial X}u_{dx2} + u_{dx2}\frac{\partial}{\partial X}u_{dx1} - \frac{\partial}{\partial X}\phi_3 = 0, \quad (3.159)$$

$$u_{dy2} = -\frac{\lambda}{\Omega}\frac{\partial u_{dz1}}{\partial X} - \frac{1}{\Omega}\frac{\partial}{\partial Z}\phi_2, \quad u_{dy2} = -\frac{\lambda}{\Omega}\frac{\partial u_{dz2}}{\partial X} - \frac{1}{\Omega}\frac{\partial}{\partial Z}\phi_3,$$

$$u_{dz2} = \frac{\lambda}{\Omega}\frac{\partial}{\partial X}u_{dy1} + \frac{1}{\Omega}\frac{\partial}{\partial Y}\phi_2, \quad u_{dz3} = \frac{\lambda}{\Omega}\frac{\partial u_{dy2}}{\partial X} + \frac{1}{\Omega}\frac{\partial \phi_3}{\partial Y}, \quad (3.160)$$

Substituting equations (3.155)-(3.156) into equation (3.141), one can get

$$\varepsilon^2 : n_{d2} = -A\phi_2 + B\phi_1^2, \quad (3.161)$$

$$\varepsilon^3 : \left(\frac{\partial^2}{\partial X^2} + \frac{\partial^2}{\partial Y^2} + \frac{\partial^2}{\partial Z^2}\right)\phi_1 - n_{d3} - A\phi_3 + 2B\phi_1\phi_2 - A_3\phi_1^3 = 0, \quad (3.162)$$

Now differentiating equation (3.162) w.r.t. X, eliminating higher order perturbed terms using the above equations and proceeding in the similar manner as done for the derivation of ZK equation (3.154), one can get the modified ZK (mZK) equation as

$$\frac{\partial \psi}{\partial \tau} + E\psi^2 \frac{\partial \psi}{\partial X} + \frac{\lambda^3}{2}\frac{\partial^3 \psi}{\partial X^3} + \frac{\lambda^3}{2}\left(1 + \frac{1}{\Omega^2 \lambda}\right)\left(\frac{\partial^3 \psi}{\partial X \partial Y^2} + \frac{\partial^3 \psi}{\partial X \partial Z^2}\right) = 0,$$

(3.163)

where $E = \frac{3}{2\lambda^3} + \frac{3B\lambda}{2} - \frac{3D\lambda^3}{2}$ and
$D = \frac{1}{48}[\mu_e(1+q_e)(3-q_e)(5-3q_e)\delta_e^3 + \mu_i(1+q_i)(3-q_i)(5-3q_i)\delta_i^3]$.

3.3 Analytical wave solutions of evolution equations

3.3.1 Analytical wave solution of the KdV equation

The KdV equation that describes small amplitude nonlinear waves in plasmas is given by

$$\frac{\partial \phi}{\partial t} + A\phi\frac{\partial \phi}{\partial x} + B\frac{\partial^3 \phi}{\partial x^3} = 0,$$

(3.164)

where A and B are coefficients of non-linear and dispersion terms, respectively. Here, ϕ, x and t are the wave profile, space variable and time, respectively.

One can use the following traveling wave transformation to find the analytical wave solutions of the KdV equation (3.164):

$$\xi = x - vt,$$

(3.165)

where v denotes velocity of traveling wave. Using the transformation (3.165) in the KdV equation (3.164), one can obtain

$$-v\frac{d\phi}{d\xi} + A\phi\frac{d\phi}{d\xi} + B\frac{d^3\phi}{d\xi^3} = 0.$$

(3.166)

Integrating above equation (3.166) with respect to ξ and using boundary conditions $\phi \to 0$, $\frac{d\phi}{d\xi} \to 0$ and $\frac{d^2\phi}{d\xi^2} \to 0$ as $\xi \to \pm\infty$, one can obtain

$$-v\phi + \frac{A}{2}\phi^2 + B\frac{d^2\phi}{d\xi^2} = 0,$$

$$\text{or,} \quad \frac{d^2\phi}{d\xi^2} = \frac{v}{B}\phi - \frac{A}{2B}\phi^2. \tag{3.167}$$

Equation (3.167) can be represented in the form of a planar dynamical system [11] as

$$\begin{cases} \frac{d\phi}{d\xi} = z, \\ \frac{dz}{d\xi} = \frac{v}{B}\phi - \frac{A}{2B}\phi^2. \end{cases} \tag{3.168}$$

The dynamical system (3.168) is a Hamiltonian system with Hamiltonian function

$$H(\phi,z) = \frac{z^2}{2} - \frac{v}{2B}\phi^2 + \frac{A}{6B}\phi^3 = h \text{ (say)}. \tag{3.169}$$

For any homoclinic orbit of the dynamical system (3.168) at $(0,0)$, one can have $H(\phi,z) = 0$, which gives

$$\frac{z^2}{2} - \frac{v}{2B}\phi^2 + \frac{A}{6B}\phi^3 = 0,$$

$$\text{or,} \quad z = \pm\sqrt{\frac{v}{B}}\,\phi\,\sqrt{1 - \frac{A}{3v}\phi}. \tag{3.170}$$

Now, from equation (3.168), one can get

$$\frac{d\phi}{d\xi} = z,$$

$$\text{or,} \quad \frac{d\phi}{d\xi} = \pm\sqrt{\frac{v}{B}}\,\phi\,\sqrt{1 - \frac{A}{3v}\phi},$$

$$\text{or,} \quad \frac{d\phi}{\phi\sqrt{1 - \frac{A}{3v}\phi}} = \pm\sqrt{\frac{v}{B}}\,d\xi. \tag{3.171}$$

One can consider a transformation,

$$\frac{A}{3v}\phi = p^2. \tag{3.172}$$

Applying the transformation (3.172) in equation (3.171), one can get

$$\frac{dp}{p\sqrt{1 - p^2}} = \pm\frac{1}{2}\sqrt{\frac{v}{B}}\,d\xi. \tag{3.173}$$

Integrating equation (3.173) and considering integration constant as 0, one can obtain

$$\int \frac{dp}{p\sqrt{1-p^2}} = \pm \int \frac{1}{2}\sqrt{\frac{v}{B}}\, d\xi,$$

$$\text{or,} \quad sech^{-1}p = \mp \frac{1}{2}\sqrt{\frac{v}{B}}\, \xi,$$

$$\text{or,} \quad p = sech\left(\mp \frac{1}{2}\sqrt{\frac{v}{B}}\, \xi\right). \tag{3.174}$$

Using transformation (3.172) in equation (3.174), one can obtain

$$\phi = \frac{3v}{A} sech^2\left(\frac{1}{2}\sqrt{\frac{v}{B}}\, \xi\right). \tag{3.175}$$

Equation (3.175) represents the solitary wave solution of the KdV equation (3.164) with amplitude $\frac{3v}{A}$ and width $2\sqrt{\frac{B}{v}}$.

3.3.2 Analytical wave solution of the mKdV equation

The nonlinear coefficient A of the KdV equation (3.164) can vanish for some critical values of the physical parameters. In such condition, modified KdV (mKdV) equation describes small amplitude nonlinear waves in such plasmas, which is given as

$$\frac{\partial \phi}{\partial t} + A\phi^2\frac{\partial \phi}{\partial x} + B\frac{\partial^3 \phi}{\partial x^3} = 0, \tag{3.176}$$

where A and B are coefficients of non-linear and dispersion terms, respectively. Here ϕ, x and t are the wave profile, space variable and time, respectively.

One can use the following traveling wave transformation to find the analytical wave solutions of the mKdV equation (3.176)

$$\xi = x - vt, \tag{3.177}$$

where v denotes velocity of traveling wave. Using the transformation (3.177) in the mKdV equation (3.176), one can obtain

$$-v\frac{d\phi}{d\xi} + A\phi^2\frac{d\phi}{d\xi} + B\frac{d^3\phi}{d\xi^3} = 0. \tag{3.178}$$

Integrating above equation (3.178) with respect to ξ and using boundary conditions $\phi \to 0, \frac{d\phi}{d\xi} \to 0$ and $\frac{d^2\phi}{d\xi^2} \to 0$ as $\xi \to \pm\infty$, one can obtain

$$-v\phi + \frac{A}{3}\phi^3 + B\frac{d^2\phi}{d\xi^2} = 0,$$

$$\text{or,} \quad \frac{d^2\phi}{d\xi^2} = \frac{v}{B}\phi - \frac{A}{3B}\phi^3. \tag{3.179}$$

Equation (3.179) can be expressed in the following planar dynamical form [12] as

$$\begin{cases} \frac{d\phi}{d\xi} = y, \\ \frac{dy}{d\xi} = \frac{v}{B}\phi - \frac{A}{3B}\phi^3. \end{cases} \tag{3.180}$$

The dynamical system (3.180) is a Hamiltonian system with Hamiltonian function

$$H(\phi, y) = \frac{y^2}{2} - \frac{v}{2B}\phi^2 + \frac{A}{12B}\phi^4. \tag{3.181}$$

For the homoclinic orbit of the dynamical system (3.180) at $(0,0)$, one can have $H(\phi, y) = 0$, which gives

$$\text{or,} \quad \frac{d\phi}{d\xi} = \pm\sqrt{\frac{v}{B}} \, \phi \, \sqrt{1 - \frac{A}{6v}\phi^2},$$

$$\text{or,} \quad \frac{d\phi}{\phi\sqrt{1 - \frac{A}{6v}\phi^2}} = \pm\sqrt{\frac{v}{B}} \, d\xi. \tag{3.182}$$

One can consider a transformation:

$$\frac{A}{6v}\phi^2 = q^2. \tag{3.183}$$

Applying equation (3.183) in equation (3.182), one can get

$$\frac{dq}{q\sqrt{1 - q^2}} = \pm\sqrt{\frac{v}{B}} \, d\xi. \tag{3.184}$$

Integrating equation (3.184) and considering integration constant as 0, one can obtain

$$\int \frac{dq}{q\sqrt{1 - q^2}} = \pm \int \sqrt{\frac{v}{B}} \, d\xi,$$

$$\text{or,} \quad sech^{-1}q = \mp \sqrt{\frac{v}{B}}\,\xi,$$

$$\text{or,} \quad q = sech\left(\mp\sqrt{\frac{v}{B}}\,\xi\right). \tag{3.185}$$

Using transformation (3.183) in equation (3.185), one can obtain

$$\phi = \pm\sqrt{\frac{6v}{A}}\,sech\left(\sqrt{\frac{v}{B}}\,\xi\right). \tag{3.186}$$

Equation (3.186) represents the compressive and rarefactive solitary wave solutions of the mKdV equation (3.176) with amplitude $\sqrt{\frac{6v}{A}}$ and width $\sqrt{\frac{B}{v}}$.

3.3.3 Analytical wave solution of the KP equation

The KP equation that describes small amplitude nonlinear waves propagating on xy-plane in plasmas is given by

$$\frac{\partial}{\partial x}\left(\frac{\partial\phi}{\partial t}+A\phi\frac{\partial\phi}{\partial x}+B\frac{\partial^3\phi}{\partial x^3}\right)+C\frac{\partial^2\phi}{\partial y^2}=0, \tag{3.187}$$

where A, B and C are coefficients of non-linear and dispersion terms, respectively. Here ϕ, x, y and t are the wave profile, space variables and time, respectively.

The following traveling wave transformation is used to find the analytical wave solutions of the KP equation (3.187)

$$\xi = lx+my-vt, \tag{3.188}$$

where v denotes velocity of traveling wave, l and m are direction cosines (DCs) of the line of propagation with respect to the x- and y-axes, respectively. Using the transformation (3.188) in the KP equation (3.187), one can obtain

$$l\frac{d}{d\xi}\left(-v\frac{d\phi}{d\xi}+Al\,\phi\frac{d\phi}{d\xi}+Bl^3\frac{d^3\phi}{d\xi^3}\right)+Cm^2\frac{d^2\phi}{d\xi^2}=0. \tag{3.189}$$

Integrating the above equation (3.189) with respect to ξ and using boundary conditions $\phi\to 0, \frac{d\phi}{d\xi}\to 0, \frac{d^2\phi}{d\xi^2}\to 0$ and $\frac{d^3\phi}{d\xi^3}\to 0$ as $\xi\to\pm\infty$, one can obtain

$$-vl\frac{d\phi}{d\xi}+Al^2\,\phi\frac{d\phi}{d\xi}+Bl^4\frac{d^3\phi}{d\xi^3}+C(1-l^2)\frac{d\phi}{d\xi}=0. \tag{3.190}$$

Integrating equation (3.190) again and using the boundary conditions, one can get

$$-vl\,\phi + \frac{Al^2}{2}\,\phi^2 + Bl^4\,\frac{d^2\phi}{d\xi^2} + C(1-l^2)\,\phi = 0,$$

$$\text{or,}\quad \frac{d^2\phi}{d\xi^2} = \frac{vl - C(1-l^2)}{Bl^4}\,\phi - \frac{A}{2Bl^2}\,\phi^2. \tag{3.191}$$

Equation (3.191) can be represented in the form of a planar dynamical system as

$$\begin{cases} \frac{d\phi}{d\xi} = z, \\ \frac{dz}{d\xi} = P\,\phi - Q\,\phi^2, \end{cases} \tag{3.192}$$

where $P = \frac{vl - C(1-l^2)}{Bl^4}$ and $Q = \frac{A}{2Bl^2}$.

The dynamical system (3.192) is a Hamiltonian system with Hamiltonian function

$$H(\phi,z) = \frac{z^2}{2} - \frac{P}{2}\,\phi^2 + \frac{Q}{3}\,\phi^3 = h \ \text{(say)}. \tag{3.193}$$

For any homoclinic orbit of the dynamical system (3.192) at $(0,0)$, one can have $H(\phi,z) = 0$, which gives

$$\frac{z^2}{2} - \frac{P}{2}\,\phi^2 + \frac{Q}{3}\,\phi^3 = 0,$$

$$\text{or,}\quad z = \pm\sqrt{P}\,\phi\,\sqrt{1 - \frac{2Q}{3P}\phi}. \tag{3.194}$$

Now, from equation (3.192), one can get

$$\frac{d\phi}{d\xi} = z,$$

$$\text{or,}\quad \frac{d\phi}{d\xi} = \pm\sqrt{P}\,\phi\,\sqrt{1 - \frac{2Q}{3P}\phi},$$

$$\text{or,}\quad \frac{d\phi}{\phi\sqrt{1 - \frac{2Q}{3P}\phi}} = \pm\sqrt{P}\,d\xi. \tag{3.195}$$

One can consider a transformation:

$$\frac{2Q}{3P}\phi = w^2. \tag{3.196}$$

Applying equation (3.196) in equation (3.195), one can get

$$\frac{dw}{w\sqrt{1-w^2}} = \pm \frac{1}{2}\sqrt{P}\, d\xi. \tag{3.197}$$

Integrating equation (3.197) and considering integration constant as 0, one can obtain

$$\int \frac{dw}{w\sqrt{1-w^2}} = \pm \int \frac{1}{2}\sqrt{P}\, d\xi,$$

$$\text{or,} \quad sech^{-1}w = \mp \frac{1}{2}\sqrt{P}\,\xi,$$

$$\text{or,} \quad w = sech\left(\mp \frac{1}{2}\sqrt{P}\,\xi\right). \tag{3.198}$$

Using transformation (3.196) in equation (3.198), one can obtain

$$\phi = \frac{3P}{2Q}sech^2\left(\frac{1}{2}\sqrt{P}\,\xi\right). \tag{3.199}$$

Thus, the solitary wave solution of the KP equation (3.187) is given by

$$\phi = \frac{3(vl - C(1 - l^2))}{Al^2}sech^2\left(\frac{1}{2}\sqrt{\frac{vl - C(1 - l^2)}{Bl^4}}\,\xi\right), \tag{3.200}$$

where, amplitude $= \dfrac{3(vl - C(1 - l^2))}{Al^2}$ and width $= 2\sqrt{\dfrac{Bl^4}{vl - C(1 - l^2)}}$.

3.3.4 Analytical wave solution of the mKP equation

The modified KP (mKP) equation that describes small amplitude nonlinear waves propagating on xy-plane in plasmas is given by

$$\frac{\partial}{\partial x}\left(\frac{\partial \phi}{\partial t} + A\phi^2\frac{\partial \phi}{\partial x} + B\frac{\partial^3 \phi}{\partial x^3}\right) + C\frac{\partial^2 \phi}{\partial y^2} = 0, \tag{3.201}$$

where A, B and C are coefficients of non-linear term and dispersion terms, respectively. Here ϕ, x, y and t are the wave profile, space variables and time, respectively.

The following traveling wave transformation is applied to find the analytical wave solutions of the mKP equation (3.201)

$$\xi = lx + my - vt, \tag{3.202}$$

where v denotes velocity of traveling wave, l and m are DCs of the line of propagation with respect to x- and y-axes, respectively.

Using the transformation (3.202) in the mKP equation (3.201), one can obtain

$$l \frac{d}{d\xi} \left(-v \frac{d\phi}{d\xi} + Al \, \phi^2 \frac{d\phi}{d\xi} + Bl^3 \frac{d^3\phi}{d\xi^3} \right) + Cm^2 \frac{d^2\phi}{d\xi^2} = 0. \tag{3.203}$$

Integrating the above equation (3.203) with respect to ξ and using boundary conditions $\phi \to 0, \frac{d\phi}{d\xi} \to 0, \frac{d^2\phi}{d\xi^2} \to 0$ and $\frac{d^3\phi}{d\xi^3} \to 0$ as $\xi \to \pm\infty$, one can obtain

$$-vl \frac{d\phi}{d\xi} + Al^2 \, \phi^2 \frac{d\phi}{d\xi} + Bl^4 \frac{d^3\phi}{d\xi^3} + C(1 - l^2) \frac{d\phi}{d\xi} = 0. \tag{3.204}$$

Integrating equation (3.204) again and using the boundary conditions, one can get

$$-vl \, \phi + \frac{Al^2}{3} \phi^3 + Bl^4 \frac{d^2\phi}{d\xi^2} + C(1 - l^2) \, \phi = 0,$$

$$\text{or,} \quad \frac{d^2\phi}{d\xi^2} = \frac{vl - C(1 - l^2)}{Bl^4} \phi - \frac{A}{3Bl^2} \phi^3. \tag{3.205}$$

Equation (3.205) can be represented in the form of a planar dynamical system as

$$\begin{cases} \frac{d\phi}{d\xi} = z, \\ \frac{dz}{d\xi} = P_1 \, \phi - Q_1 \, \phi^3, \end{cases} \tag{3.206}$$

where $P_1 = \frac{vl - C(1 - l^2)}{Bl^4}$ and $Q_1 = \frac{A}{3Bl^2}$.

The dynamical system (3.206) is a Hamiltonian system with Hamiltonian function

$$H(\phi, z) = \frac{z^2}{2} - \frac{P_1}{2} \phi^2 + \frac{Q_1}{4} \phi^4 = h \text{ (say)}. \tag{3.207}$$

For the homoclinic orbit of the dynamical system (3.206) at $(0,0)$, one can have $H(\phi, z) = 0$, which gives

$$\frac{z^2}{2} - \frac{P_1}{2} \phi^2 + \frac{Q_1}{4} \phi^4 = 0,$$

$$\text{or,} \quad z = \pm\sqrt{P_1}\,\phi\,\sqrt{1 - \frac{Q_1}{2P_1}\phi^2}. \tag{3.208}$$

Now, from equation (3.206), one can get

$$\frac{d\phi}{d\xi} = z,$$

$$\text{or,} \quad \frac{d\phi}{d\xi} = \pm\sqrt{P_1}\,\phi\,\sqrt{1 - \frac{Q_1}{2P_1}\phi^2},$$

$$\text{or,} \quad \frac{d\phi}{\phi\sqrt{1 - \frac{Q_1}{2P_1}\phi^2}} = \pm\sqrt{P_1}\,d\xi. \tag{3.209}$$

One can consider a transformation:

$$\frac{Q_1}{2P_1}\phi^2 = w^2. \tag{3.210}$$

Applying the transformation (3.210) in equation (3.209), one can obtain

$$\frac{dw}{w\sqrt{1 - w^2}} = \pm\sqrt{P_1}\,d\xi. \tag{3.211}$$

Integrating equation (3.211) and considering integration constant as 0, one can obtain

$$\int \frac{dw}{w\sqrt{1 - w^2}} = \pm \int \sqrt{P_1}\,d\xi,$$

$$\text{or,} \quad sech^{-1}w = \mp\sqrt{P_1}\,\xi,$$

$$\text{or,} \quad w = sech\left(\mp\sqrt{P_1}\,\xi\right). \tag{3.212}$$

Using transformation (3.210) in equation (3.212), one can obtain

$$\phi = \pm\sqrt{\frac{2P_1}{Q_1}}\,sech\left(\sqrt{P_1}\,\xi\right). \tag{3.213}$$

Thus, the compressive and rarefactive solitary wave solutions of the mKP equation (3.201) are given by

$$\phi = \pm\frac{6(vl - C(1 - l^2))}{Al^2}\,sech\left(\sqrt{\frac{vl - C(1 - l^2)}{Bl^4}}\,\xi\right), \tag{3.214}$$

where, amplitude $= \dfrac{6(vl - C(1 - l^2))}{Al^2}$ and width $= \sqrt{\dfrac{Bl^4}{vl - C(1 - l^2)}}$.

3.3.5 Analytical wave solution of the ZK equation

The ZK equation defines the feature of weakly nonlinear waves in plasmas. The standard form of the ZK equation in three-dimensional space is given by:

$$\frac{\partial \phi}{\partial t} + A\phi \frac{\partial \phi}{\partial x} + B \frac{\partial}{\partial x} \left(\frac{\partial^2 \phi}{\partial x^2} + \frac{\partial^2 \phi}{\partial y^2} + \frac{\partial^2 \phi}{\partial z^2} \right) = 0, \tag{3.215}$$

where A and B are coefficients of non-linear and dispersion terms, respectively. Here ϕ, (x, y, z) and t are the wave profile, space variables and time, respectively.

One can consider the following traveling wave transformation to find the analytical wave solution of the ZK equation (3.215)

$$\xi = lx + my + nz - vt, \tag{3.216}$$

where l, m and n are the D.Cs. of line of flow with velocity v. Using the transformation (3.216) in the ZK equation (3.215), one can get

$$-v\frac{d\phi}{d\xi} + Al\phi\frac{d\phi}{d\xi} + Bl\frac{d}{d\xi}\left(l^2\frac{d^2\phi}{d\xi^2} + m^2\frac{d^2\phi}{d\xi^2} + n^2\frac{d^2\phi}{d\xi^2}\right) = 0,$$

$$\text{or,} \quad -v\frac{d\phi}{d\xi} + Al\phi\frac{d\phi}{d\xi} + Bl\frac{d^3\phi}{d\xi^3} = 0, \ (\because \ l^2 + m^2 + n^2 = 1) \tag{3.217}$$

Integrating equation (3.217) with respect to ξ and using boundary conditions $\phi \to 0, \frac{d\phi}{d\xi} \to 0$ and $\frac{d^2\phi}{d\xi^2} \to 0$ as $\xi \to \pm\infty$, one can get

$$-v\phi + \frac{Al}{2}\phi^2 + Bl\frac{d^2\phi}{d\xi^2} = 0,$$

$$\text{or,} \quad \frac{d^2\phi}{d\xi^2} = \frac{v}{Bl}\phi - \frac{A}{2B}\phi^2. \tag{3.218}$$

Writing equation (3.218) in the form of a planar dynamical system [12], one can obtain

$$\begin{cases} \frac{d\phi}{d\xi} = \psi, \\ \frac{d\psi}{d\xi} = \frac{v}{Bl}\phi - \frac{A}{2B}\phi^2. \end{cases} \tag{3.219}$$

The dynamical system (3.219) is a Hamiltonian system with Hamiltonian function

$$H(\phi, \psi) = \frac{\psi^2}{2} - \frac{v}{2Bl}\phi^2 + \frac{A}{6B}\phi^3. \tag{3.220}$$

For any homoclinic orbit of the dynamical system (3.219) at $(0,0)$, one can have $H(\phi, \psi) = 0$, which gives

$$\frac{\psi^2}{2} - \frac{v}{2Bl}\phi^2 + \frac{A}{6B}\phi^3 = 0,$$

$$\text{or,} \quad \psi = \pm \sqrt{\frac{v}{Bl}} \, \phi \, \sqrt{1 - \frac{Al}{3v}\phi}. \tag{3.221}$$

Now, from equation (3.219), one can get

$$\frac{d\phi}{d\xi} = \psi,$$

$$\text{or,} \quad \frac{d\phi}{d\xi} = \pm \sqrt{\frac{v}{Bl}} \, \phi \, \sqrt{1 - \frac{Al}{3v}\phi},$$

$$\text{or,} \quad \frac{d\phi}{\phi\sqrt{1 - \frac{Al}{3v}\phi}} = \pm \sqrt{\frac{v}{Bl}} \, d\xi. \tag{3.222}$$

one can take a transformation:

$$\frac{Al}{3v}\phi = p^2. \tag{3.223}$$

Applying the transformation (3.223) in equation (3.222), one can get

$$\frac{dp}{p\sqrt{1 - p^2}} = \pm \frac{1}{2}\sqrt{\frac{v}{Bl}} \, d\xi. \tag{3.224}$$

Integrating equation (3.224) and considering integration constant as 0, one can obtain

$$\int \frac{dp}{p\sqrt{1 - p^2}} = \pm \int \frac{1}{2}\sqrt{\frac{v}{Bl}} \, d\xi,$$

$$\text{or,} \quad \operatorname{sech}^{-1} p = \mp \frac{1}{2}\sqrt{\frac{v}{Bl}} \, \xi,$$

$$\text{or,} \quad p = \operatorname{sech}\left(\mp \frac{1}{2}\sqrt{\frac{v}{Bl}} \, \xi\right). \tag{3.225}$$

Using transformation (3.223) in equation (3.225), one can obtain

$$\phi = \frac{3v}{Al} \, sech^2 \left(\frac{1}{2} \sqrt{\frac{v}{Bl}} \, \xi \right). \tag{3.226}$$

Equation (3.226) is the solitary wave solution of the ZK equation (3.215) with amplitude $\frac{3v}{Al}$ and width $2\sqrt{\frac{Bl}{v}}$.

It is important to note that if $A \equiv 0$, then the ZK equation (3.215) is unable to describe the wave feature given by equation (3.226). So, in this situation, one needs to consider higher order correction and the modified ZK equation can describe the nonlinear wave features. Thus, the analytical forms of such nonlinear wave features are discussed in next subsection.

3.3.6 Analytical wave solution of the mZK equation

The modified ZK (mZK) equation in three dimensional space is given by

$$\frac{\partial \phi}{\partial t} + A\phi^2 \frac{\partial \phi}{\partial x} + B \frac{\partial}{\partial x} \left(\frac{\partial^2 \phi}{\partial x^2} + \frac{\partial^2 \phi}{\partial y^2} + \frac{\partial^2 \phi}{\partial z^2} \right) = 0, \tag{3.227}$$

where A and B are coefficients of non-linear and dispersion terms, respectively. Here ϕ, (x,y,z) and t are the wave profile, space variables and time, respectively.

One can take the following traveling wave transformation to find the solitary wave feature of the mZK equation (3.227)

$$\xi = lx + my + nz - vt, \tag{3.228}$$

where l, m and n are the D.Cs. of line of flow with velocity v. Using the transformation (3.228) in the mZK equation (3.227), one can get

$$-v \frac{d\phi}{d\xi} + Al\phi^2 \frac{d\phi}{d\xi} + Bl \frac{d}{d\xi} \left(l^2 \frac{d^2\phi}{d\xi^2} + m^2 \frac{d^2\phi}{d\xi^2} + n^2 \frac{d^2\phi}{d\xi^2} \right) = 0,$$

or, $\quad -v \frac{d\phi}{d\xi} + Al\phi^2 \frac{d\phi}{d\xi} + Bl \frac{d^3\phi}{d\xi^3} = 0, \quad (\because \ l^2 + m^2 + n^2 = 1) \tag{3.229}$

Integrating equation (3.229) over ξ and using boundary conditions $\phi \to 0, \frac{d\phi}{d\xi} \to 0$ and $\frac{d^2\phi}{d\xi^2} \to 0$ as $\xi \to \pm\infty$, one can obtain

$$-v\phi + \frac{Al}{3}\phi^3 + Bl\frac{d^2\phi}{d\xi^2} = 0,$$

or, $\qquad \frac{d^2\phi}{d\xi^2} = \frac{v}{Bl}\phi - \frac{A}{3B}\phi^3. \qquad (3.230)$

Expressing above equation (3.230) in the form of dynamical system [11], one can get

$$\begin{cases} \frac{d\phi}{d\xi} = \psi, \\ \frac{d\psi}{d\xi} = \frac{v}{Bl}\phi - \frac{A}{3B}\phi^3. \end{cases} \qquad (3.231)$$

The dynamical system (3.231) is a Hamiltonian system with Hamiltonian function

$$H(\phi, \psi) = \frac{\psi^2}{2} - \frac{v}{2Bl}\phi^2 + \frac{A}{12B}\phi^4. \qquad (3.232)$$

For homoclinic orbit of the dynamical system (3.231) at $(0,0)$, one can have $H(\phi, \psi) = 0$, which gives

$$\frac{\psi^2}{2} - \frac{v}{2Bl}\phi^2 + \frac{A}{12B}\phi^4 = 0,$$

or, $\qquad \psi = \pm\sqrt{\frac{v}{Bl}}\,\phi\,\sqrt{1 - \frac{Al}{6v}\phi^2}. \qquad (3.233)$

Now, from equation (3.231), one can have

$$\frac{d\phi}{d\xi} = \psi,$$

or, $\qquad \frac{d\phi}{d\xi} = \pm\sqrt{\frac{v}{Bl}}\,\phi\,\sqrt{1 - \frac{Al}{6v}\phi^2},$

or, $\qquad \dfrac{d\phi}{\phi\sqrt{1 - \frac{Al}{6v}\phi^2}} = \pm\sqrt{\frac{v}{Bl}}\,d\xi. \qquad (3.234)$

One can take a transformation:

$$\frac{Al}{6v}\phi^2 = p^2. \qquad (3.235)$$

Applying the transformation (3.235) in equation (3.234), one can get

$$\frac{dp}{p\sqrt{1-p^2}} = \pm\sqrt{\frac{v}{Bl}}\, d\xi. \tag{3.236}$$

Integrating equation (3.236) and considering integration constant as 0, one can obtain

$$\int \frac{dp}{p\sqrt{1-p^2}} = \pm\int \sqrt{\frac{v}{Bl}}\, d\xi,$$

$$\text{or,} \quad sech^{-1}p = \mp\sqrt{\frac{v}{Bl}}\,\xi,$$

$$\text{or,} \quad p = sech\left(\mp\sqrt{\frac{v}{Bl}}\,\xi\right). \tag{3.237}$$

Using transformation (3.235) in equation (3.237), one can obtain

$$\phi = \pm\sqrt{\frac{6v}{Al}}\, sech\left(\sqrt{\frac{v}{Bl}}\,\xi\right). \tag{3.238}$$

Equation (3.238) provides the solitary wave solutions of the mZK equation (3.227) with amplitude $\sqrt{\frac{6v}{Al}}$ and width $\sqrt{\frac{Bl}{v}}$.

3.3.7 Analytical wave solution of the Burgers equation

The Burgers equation for small amplitude nonlinear wave in plasmas with viscosity effect is given by

$$\frac{\partial\psi}{\partial t} + A\psi\frac{\partial\psi}{\partial x} - B\frac{\partial^2\psi}{\partial x^2} = 0, \tag{3.239}$$

where A is the coefficient of nonlinear term and $B = \frac{\eta}{2}$, here η is the coefficient of kinematic viscosity.

To find the traveling wave solution of the Burgers equation (3.239), one can consider the Galilean transformation

$$\xi = x - vt, \tag{3.240}$$

where v is traveling wave velocity.

Then, equation (3.239) becomes

$$-v\frac{d\psi}{d\xi} + A\psi\frac{d\psi}{d\xi} - B\frac{d^2\psi}{d\xi^2} = 0. \tag{3.241}$$

Integrating equation (3.241) w.r.t. ξ and applying boundary conditions $\psi \to 0$, $\frac{d\psi}{d\xi} \to 0$ as $\xi \to \pm\infty$, one can have

$$-v\psi + A\frac{\psi^2}{2} - B\frac{d\psi}{d\xi} = 0,$$

or, $$\frac{d\psi}{d\xi} = -\frac{v}{B}\psi + \frac{A}{2B}\psi^2,$$

or, $$\frac{d\psi}{d\xi} = \frac{A}{2B}\left(\psi^2 - \frac{2v}{A}\psi\right),$$

or, $$\frac{d\psi}{d\xi} = \frac{A}{2B}\left(\psi^2 - \frac{2v}{A}\psi + \frac{v^2}{A^2} - \frac{v^2}{A^2}\right),$$

or, $$\frac{d\psi}{d\xi} = \frac{A}{2B}\left[\left(\psi - \frac{v}{A}\right)^2 - \frac{v^2}{A^2}\right],$$

$$\tag{3.242}$$

or, $$\frac{d\psi}{\left[\left(\psi - \frac{v}{A}\right)^2 - \frac{v^2}{A^2}\right]} = \frac{A}{2B}d\xi, \tag{3.243}$$

or, $$\frac{d\left(\psi - \frac{v}{A}\right)}{\left[\frac{v^2}{A^2} - \left(\psi - \frac{v}{A}\right)^2\right]} = -\frac{A}{2B}d\xi. \tag{3.244}$$

Integrating equation (3.244), one can get

$$\frac{A}{v}\tanh^{-1}\left(\frac{A\psi - v}{v}\right) = -\frac{A}{2B}\xi + c_1, \tag{3.245}$$

where c_1 is an integral constant. Taking $c_1 = 0$, one can get

$$\frac{A}{v} tanh^{-1} \left(\frac{A\psi - v}{v} \right) = -\frac{A}{2B} \xi,$$

$$\text{or,} \quad \left(\frac{A\psi - v}{v} \right) = tanh \left(-\frac{v}{2B} \xi \right),$$

$$\text{or,} \quad \left(\frac{A\psi - v}{v} \right) = -tanh \left(\frac{v}{2B} \xi \right),$$

$$\text{or,} \quad \psi = \left(\frac{v}{A} \right) \left[1 - tanh \left(\frac{v}{2B} \xi \right) \right],$$

$$\text{or,} \quad \psi = \psi_m \left[1 - tanh \left(\frac{\xi}{\delta} \right) \right], \tag{3.246}$$

where $\psi_m = \dfrac{v}{A}$ and $\delta = \dfrac{2B}{v}$.

Equation (3.246) is the required kink or anti-kink wave solution of the Burgers equation (3.239).

References

[1] F. F. Chen, Introduction to Plasma Physics and Controlled Fusion, volume 1, Plenum Press, Second edition (1984).

[2] J. A. Bittencourt, Fundamentals of Plasma Physics, Pergamon Press, First edition (1986).

[3] N. N. Rao, P. K. Shukla and M. Y. Yu, Planet Space Sci., 38: 543 (1990).

[4] A. Barken, N. D'Angelo and R. L. Merlino, Planet Space Sci., 43: 905 (1995).

[5] J. B. Pieper and J. Goree, Phys. Rev. Lett., 77: 3137 (1996).

[6] P. K. Shukla and V. P. Silin, Physica Scripta, 45: 508 (1992).

[7] A. Barken, N. D'Angelo and R. L. Merlino, Planet Space Sci., 44: 239 (1996).

[8] Y. Nakamura, B. Bailung and P. K. Shukla, Phys. Rev. Lett., 83: 6 (1999).

[9] A. S. Bains, M. Tribeche and T. S. Gill, Phys. Plasmas, 18: 022108 (2011).

[10] W. M. Moslem, Phys. Plasmas, 10: 3168 (2003).

[11] S. H. Strogatz, Nonlinear Dynamics and Chaos, Westview Press (USA) (2007).

[12] G. C. Layek, An Introduction to Dynamical Systems and Chaos, Springer, India (2015).

Chapter 4

Bifurcation of Small Amplitude Waves in Plasmas

4.1 Introduction

A nonlinear wave with amplitude less than unity is called small amplitude nonlinear wave. Reductive perturbation technique (RPT) plays a crucial role in the investigation of small-amplitude nonlinear waves. Mathematically, RPT redefines space and timescale [1] in fundamental model equations of systems that describe long wavelength situation. Applying RPT, governing equations are simplified to nonlinear evolution equations, such as the Burgers equation, the KdV equation, etc. Nonlinear acoustic wave phenomena are studied employing RPT by some researchers [2]-[3]. Therefore, it is noteworthy that RPT may be employed to study small-amplitude nonlinear waves in plasmas. Now, applying the bifurcation theory of planar dynamical systems (DSs), one can study bifurcation behaviour of small amplitude nonlinear waves in various plasma systems. There are different kind of small amplitude nonlinear waves in plasmas, for example,

- Small amplitude ion-acoustic wave.

- Small amplitude dust-ion-acoustic wave.

■ Small amplitude dust-acoustic wave.

■ Small amplitude electron-acoustic wave, etc.

4.2 Bifurcation of ion-acoustic waves with small amplitude

Firstly, bifurcation behaviour of ion-acoustic waves (IAWs) with small amplitude is presented in a three-component unmagnetised plasma consisting of cold mobile ions and κ-distributed cold and hot electrons. In this case, employing RPT, the KdV equation is presented to discuss possible wave solutions.

4.2.1 Basic equations

Small-amplitude IAWs in a three-component plasma system consisting of cold fluid ions and kappa distributed electrons of different temperatures (hot and cold) are examined. Propagation of IAWs is governed by the following basic equations [4]:

$$\frac{\partial n}{\partial t} + \frac{\partial}{\partial x}(nu) = 0, \qquad (4.1)$$

$$\frac{\partial u}{\partial t} + u\frac{\partial u}{\partial x} = -\frac{\partial \phi}{\partial x}, \qquad (4.2)$$

$$\frac{\partial^2 \phi}{\partial x^2} = f\left(1 - \frac{\alpha_c \phi}{\kappa - \frac{3}{2}}\right)^{-\kappa + \frac{1}{2}} + (1-f)\left(1 - \frac{\alpha_h \phi}{\kappa - \frac{3}{2}}\right)^{-\kappa + \frac{1}{2}} - n, \qquad (4.3)$$

where n, ϕ, u and f are number density of cold ions, electrostatic potential, velocity of ions and fractional charge density of cold electrons, respectively. Here, $\alpha_c = \frac{T_{eff}}{T_c}$ and $\alpha_h = \frac{T_{eff}}{T_h}$, where $T_{eff} = \frac{T_c T_h}{f T_h + (1-f) T_c}$ is effective temperature with cold and hot electron temperatures T_c and T_h, respectively, [5].

The considered plasma system is normalized by: n_0 normalizes n, $C_s = \left(\frac{k_B T_e}{m}\right)^{\frac{1}{2}}$ normalizes u, where k_B stands for the Boltzmann constant, m denotes ion mass, and e stands for strength of electron charge. Here, $\frac{k_B T_e}{e}$ normalizes ϕ, $\omega^{-1} = \left(\frac{m}{4\pi n_0 e^2}\right)^{\frac{1}{2}}$ normalizes t, where ω depicts frequency of plasma and the Debye length $\lambda_D = \left(\frac{k_B T_e}{4\pi n_0 e^2}\right)^{\frac{1}{2}}$ normalizes x.

4.2.2 Derivation of the KdV equation

The KdV equation can be derived using the following stretching

$$\xi = \varepsilon^{\frac{1}{2}}(x - vt) \quad \text{and} \quad \tau = \varepsilon^{\frac{3}{2}}t, \qquad (4.4)$$

where ε measures weakness of nonlinearity and v denotes phase velocity of IAWs. Expansions for dependent variables are:

$$\begin{cases} n = 1 + \varepsilon n_1 + \varepsilon^2 n_2 + \varepsilon^3 n_3 \dots \\ u = 0 + \varepsilon u_1 + \varepsilon^2 u_2 + \varepsilon^3 u_3 \dots \\ \phi = 0 + \varepsilon \phi_1 + \varepsilon^2 \phi_2 + \varepsilon^3 \phi_3 \dots \end{cases} \qquad (4.5)$$

Substituting equations (4.4)-(4.5) in fundamental equations (4.1)-(4.3), one can obtain the following relations comparing the coefficients of $\varepsilon^{3/2}$

$$n_1 = \frac{1}{v}u_1, \quad u_1 = \frac{1}{v}\phi_1, \qquad (4.6)$$

$$v^2 = \frac{1}{a(f\alpha_c + (1-f)\alpha_h)}, \qquad (4.7)$$

where $a = \frac{\kappa - \frac{1}{2}}{\kappa - \frac{3}{2}}$. Comparing the coefficients of $\varepsilon^{5/2}$, the following equations are obtained

$$\frac{\partial n_1}{\partial \tau} - v\frac{\partial n_2}{\partial \xi} + \frac{\partial}{\partial \xi}(n_1 u_1) + \frac{\partial}{\partial \xi}u_2 = 0, \quad (4.8)$$

$$\frac{\partial u_1}{\partial \tau} - v\frac{\partial u_2}{\partial \xi} + u_1\frac{\partial u_1}{\partial \xi} + \frac{\partial \phi_2}{\partial \xi} = 0, \quad (4.9)$$

$$\frac{\partial^2 \phi_1}{\partial \xi^2} + n_2 - a\phi_2(f\alpha_c + (1-f)\alpha_h) - b\phi_1^2(f\alpha_c^2 + (1-f)\alpha_h^2) = 0, \quad (4.10)$$

where $b = \frac{\kappa^2 - \frac{1}{4}}{2(\kappa - \frac{3}{2})^2}$. Differentiating equation (4.10) w.r. to ξ and eliminating all higher order perturbed terms using equations (4.6)-(4.9), the KdV equation is derived as

$$\frac{\partial \phi_1}{\partial \tau} + A\phi_1\frac{\partial \phi_1}{\partial \xi} + B\frac{\partial^3 \phi_1}{\partial \xi^3} = 0, \qquad (4.11)$$

where $A = \frac{v^3}{2}(\frac{3}{v^4} - 2b(f\alpha_c^2 + (1-f)\alpha_h^2))$ and $B = \frac{v^3}{2}$.

4.2.3 Formation of dynamical system

One can consider $\chi = \xi - V\tau$ as traveling wave transformation, where velocity of traveling wave is denoted by V. Then, equation (4.11) becomes

$$-V\frac{d\phi_1}{d\chi} + A\phi_1\frac{d\phi_1}{d\chi} + B\frac{d^3\phi_1}{d\chi^3} = 0. \qquad (4.12)$$

Integrating equation (4.12) w.r. to χ and using conditions $\phi_1 \to 0$, $\frac{d\phi_1}{d\chi} \to 0$, $\frac{d\phi_1}{d\chi} \to 0$ as $\chi\pm \to \infty$, one gets

$$\frac{d^2\phi_1}{d\chi^2} = \frac{1}{B}\left(V\phi_1 - \frac{A}{2}\phi_1^2\right). \qquad (4.13)$$

Then, equation (4.13) is presented in the form of a DS as:

$$\begin{cases} \frac{d\Phi}{d\chi} = z, \\ \frac{dz}{d\chi} = \frac{1}{B}\left(V\Phi - \frac{A}{2}\Phi^2\right), \end{cases} \qquad (4.14)$$

where $\Phi = \phi_1$.

4.2.4 Phase plane analysis

Mathematically, bifurcation in DSs [6] occurs when a small variation in physical parameters of that system causes qualitative change in that system. The fundamental feature of DSs is given by phase plane analysis of the bifurcation theory. It is reported that solution of traveling wave corresponds to qualitative orbit in phase plane [7]. Through the bifurcation analysis of DSs, qualitatively different phase profiles are depicted for system (4.14) with parameters κ, f, α_c and α_h. Critical points of the system (4.14) are acquired by solving the following equations simultaneously

$$\frac{d\Phi}{d\chi} = 0 \text{ and } \frac{dz}{d\chi} = 0,$$

which give

$$z = 0 \text{ and } \Phi\frac{1}{B}(V - \frac{A}{2}\Phi) = 0,$$

$$\Rightarrow z = 0 \text{ and } \Phi = 0, \frac{2V}{A}.$$

Clearly one can observe that the system (4.14) has two critical points $E_0(\Phi_0,0)$ and $E_1(\Phi_1,0)$, where $\Phi_0 = 0$ and $\Phi_1 = \frac{2V}{A}$. The Jacobian matrix J for any system

$$\begin{cases} \frac{d\Phi}{d\chi} = f(\Phi,z), \\ \frac{dz}{d\chi} = g(\Phi,z), \end{cases}$$

is given by

$$J = \begin{pmatrix} \frac{\partial f(\Phi,z)}{\partial \Phi} & \frac{\partial f(\Phi,z)}{\partial z} \\ \frac{\partial g(\Phi,z)}{\partial \Phi} & \frac{\partial g(\Phi,z)}{\partial z} \end{pmatrix}. \tag{4.15}$$

Employing equation (4.15), the Jacobian matrix of the system (4.14) at $(\Phi_i,0)$ is

$$J = \begin{pmatrix} 0 & 1 \\ \frac{1}{B}(V - A\Phi_i) & 0 \end{pmatrix},$$

and determinant of J is expressed by M as

$$M = det J(\Phi_i,0) = -\frac{1}{B}(V - A\Phi_i),$$

where $i = 0,1$. If $M < 0$, then critical point $E_i(\Phi_i,0)$ is a saddle node and for $M > 0$, critical point $E_i(\Phi_i,0)$ is a center [6]. With support of numerical computations, phase portrait profiles for the system (4.14) are presented in Figures 4.1 and 4.2, depending on system parameters κ and α_h with fixed values of α_c, f and V.

In Figure 4.1, phase profile for the KdV equation (4.11) is displayed for $\kappa = 1.6$, $\alpha_h = 0.1$, $\alpha_c = 1.1$, $f = 0.1$, and $V = 0.9$. It is observed that there exists saddle point at $E_0(0,0)$ and center at $E_1(\phi_1,0)$. A nonlinear homoclinic orbit (NHO$_{1,0}$) about $E_0(0,0)$ enclosing fixed point $E_1(\phi_1,0)$ and nonlinear periodic orbit (NPO$_{1,0}$) enclosing fixed point $E_1(\phi_1,0)$ corresponds to IASW and NPIAW solutions, respectively.

Similarly, in Figure 4.2, phase profile for the KdV equation (4.11) is presented for $\kappa = 1.7$, $\alpha_h = 0.3$, $\alpha_c = 1.1$, $f = 0.01$ and $V = 0.2$. Nonlinear homoclinic orbit (NHO$_{1,0}$) starting and terminating at $E_0(0,0)$ enclosing fixed point $E_1(\phi_1,0)$ corresponds to IASW solution and nonlinear periodic orbit (NPO$_{1,0}$) enclosing fixed point $E_1(\phi_1,0)$ corresponds to NPIAW solution.

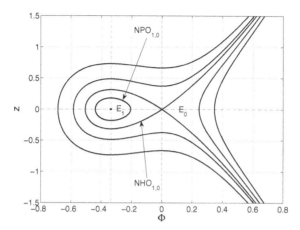

Figure 4.1: Phase profile of DS (4.14) with $\kappa = 1.6$, $\alpha_h = 0.1$, $\alpha_c = 1.1$, $f = 0.1$, and $V = 0.4$.

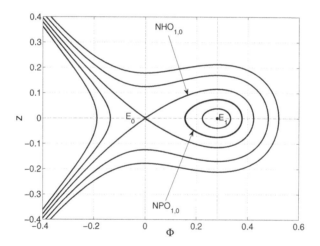

Figure 4.2: Phase profile of DS (4.14) for $\kappa = 1.7$, $\alpha_h = 0.3$, $\alpha_c = 1.1$, $f = 0.01$ and $V = 0.1$.

4.2.5 Wave solutions

The existence of ion-acoustic solitary wave solutions (IASWS) is encountered for the KdV equation from their respective phase profiles Figures (4.1)-(4.2). Hence, the analytical IASW solution for the KdV equation is obtained as:

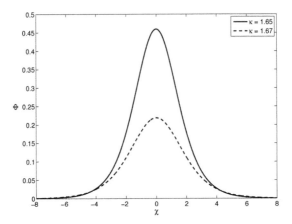

Figure 4.3: IASW solution of the KdV equation (4.11) for separate parameter values of κ with $\alpha_h = 0.1$, $\alpha_c = 1.1$, $f = 0.1$ and $V = 0.9$.

The KdV equation (4.11) supports IASW solution given by:

$$\Phi = \frac{3P}{2S} sech^2 \left(\sqrt{\frac{P}{4}} \chi \right), \qquad (4.16)$$

where $P = \frac{(\kappa - \frac{1}{2})}{(\kappa - \frac{3}{2})}(f\alpha_c + (1-f)\alpha_h) - \frac{1}{v^2}$, $S = \frac{3}{2v^4} - \frac{(\kappa^2 - \frac{1}{4})}{2(\kappa - \frac{3}{2})^2}(f\alpha_c^2 + (1-f)\alpha_h^2)$, where amplitude is $\frac{3P}{2S}$ and width is $\sqrt{\frac{4}{P}}$.

In Figure 4.3, one can notice change in IASWs of the KdV equation (4.11) with discrete values of κ and other system parameter values as $\alpha_h = 0.1$, $\alpha_c = 1.1$, $f = 0.1$ and $V = 0.9$. From Figure 4.3, it is apparent that when electrons advance away from Maxwellian, IASWs become spiky. As a result, increase in spectral index (κ) of electrons shows decrease in amplitude and increase in width of IASWs .

In Figure 4.4, change in IASWs of the KdV equation (4.11) for separate values of α_h with $\kappa = 1.7$, $\alpha_c = 1.1$, $f = 0.1$ and $V = 0.9$ is displayed. Clearly, one can notice that if temperature of hot electrons increases, α_h decreases. As a result, amplitude of IASW decreases and width increases. So, IASW becomes smooth.

Effects of κ and α on NPIAW of the KdV equation (4.11) are shown in Figures 4.5 and 4.6. When spectral index (κ) of electrons is raised, both am-

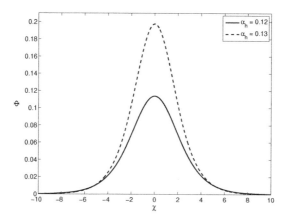

Figure 4.4: IASW solution of the KdV equation (4.11) for distinct parameter values of α_h with $\kappa = 1.7$, $\alpha_c = 1.1$, $f = 0.1$ and $V = 0.9$.

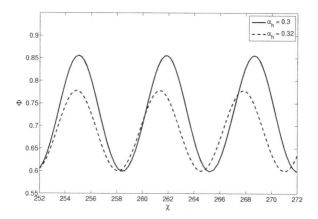

Figure 4.5: NPIAW solution of the KdV equation (4.11) for distinct parameter values of κ in with $\alpha_c = 1.1$, $f = 0.01$ and $V = 0.2$.

plitude as well as width of NPIAW are elevated. Alternately, when temperature of hot electrons is enhanced, α_h decreases and results in dwindling of both amplitude and width of NPIAW.

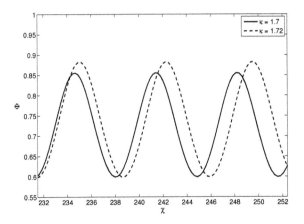

Figure 4.6: NPIAW solution of the KdV equation (4.11) for different values of α_h in with $\alpha_c = 1.1$, $f = 0.01$ and $V = 0.2$.

4.3 Bifurcation of dust-ion-acoustic waves with small amplitude

Dusty Plasma can be accounted as an ionized gas which contains charged dust particles. It is found in astrophysical objects (e.g., in the radial structure of Saturn's rings) [8]-[9]. Moreover, it has applications in laboratory and technological findings, such as in plasma discharge [10], low temperature physics (radio frequency) and in fabrication of many modern materials such as optical fibres, semi conductors, dusty crystals [11]-[13], etc. It is confirmed that charged dust grain in plasma brings up a number of new wave modes, such as dust ion acoustic (DIA) mode [14]-[15], dust acoustic (DA) mode [16]-[17], dust-drift mode [18], dust lattice (DL) mode[19] and Shukla-Varma mode [20], etc. Dust ion acoustic waves (DIAW) and dust acoustic waves (DAW) were also observed experimentally [21].

4.3.1 Governing equations

The governing equations are as follows:

$$\frac{\partial n_i}{\partial t} + \nabla.(n_i v_i) = 0, \tag{4.17}$$

$$\frac{\partial v_i}{\partial t} + (v_i.\nabla)v_i = -\frac{e\nabla\phi}{m_i} + \frac{eB_0}{m_i c}v_i \times e_z, \tag{4.18}$$

$$\nabla^2\phi = -4\pi[-en_e + en_i - ez_d n_d], \tag{4.19}$$

where n_e, n_i and n_d are number densities of electrons, ions and dusts, respectively. Here v_i, ϕ and m_i are the velocity, electrostatic potential and mass of ions, respectively. Here, z_d represents dust charge number and charge of the dust can be expressed as $q_d = -ez_d$, where e is the elementary charge. In order to observe features of electron nonextensivity, one can employ one dimensional equilibrium q-distribution function [22] given by

$$f_e(v_x) = C_q\{1 - (q-1)[\frac{m_e v_x^2}{2T_e} - \frac{e\phi}{T_e}]\}^{\frac{1}{(q-1)}}. \tag{4.20}$$

The normalized constant is

$$C_q = n_{e0}\frac{\Gamma(1/1-q)}{\Gamma(\frac{1}{1-q}-\frac{1}{2})}\sqrt{\frac{m_e(1-q)}{2\pi T_e}}, -1 < q < 1 \tag{4.21}$$

$$= n_{e0}\frac{1+q}{2}\frac{\Gamma(\frac{1}{q-1}+\frac{1}{2})}{\Gamma(1/q-1)}\sqrt{\frac{m_e(q-1)}{2\pi T_e}}, q > 1, \tag{4.22}$$

where the parameter q represents strength of nonextensivity. It is important to note that $q < -1$ and the q-distribution is unnormalizable. For $q=1$ (i.e., extensive limit) it reduces to Maxwell-Boltzman velocity distribution. After integrating $f_e(v_x)$ it takes the form

$$n_e(\phi) = n_{e0}\{1 + (q-1)\frac{e\phi}{T_e}\}^{1/(q+1)+1/2}. \tag{4.23}$$

The propagation of wave is in the xz-plane. Following normalization, the system reduces to

$$\frac{\partial n}{\partial t} + \frac{\partial(nv_x)}{\partial x} + \frac{\partial(nv_z)}{\partial z} = 0, \tag{4.24}$$

$$\frac{\partial v_x}{\partial t} + (v_x\frac{\partial}{\partial x} + v_z\frac{\partial}{\partial z})v_x = -\frac{\partial\phi}{\partial x} + v_y, \tag{4.25}$$

$$\frac{\partial v_y}{\partial t} + (v_x\frac{\partial}{\partial x} + v_z\frac{\partial}{\partial z})v_y = -v_x, \tag{4.26}$$

$$\frac{\partial v_z}{\partial t} + (v_x\frac{\partial}{\partial x} + v_z\frac{\partial}{\partial z})v_z = -\frac{\partial\phi}{\partial z}, \tag{4.27}$$

$$(\frac{\partial^2}{\partial x^2} + \frac{\partial^2}{\partial z^2})\phi = \beta[(1 + (q-1)\phi)^{\frac{q+1}{2(q-1)}} - \delta_1 n + \delta_2], \tag{4.28}$$

where $\beta = \dfrac{r_g^2}{\lambda_e^2}$, $\delta_1 = \dfrac{n_{i_0}}{n_{e_0}}$ and $\delta_2 = \dfrac{n_d z_d}{n_{e_0}}$. Here, $r_g = \dfrac{C_s}{\Omega}$ represents the

gyroradius of ions and $\lambda_e = \left(\dfrac{T_e}{4\pi n_{e_0} e^2}\right)^{1/2}$ is the Debye length of electrons.

The normalizations are as follows: $\Omega t \to t$, $(C_s/\Omega)\nabla \to \nabla$, $v_i/C_s \to v$, $n_i/n_{i_0} \to n$, $e\phi/T_e \to \phi$, where $C_s = (T_e/m_i)^{1/2}$ is the ion acoustic velocity and $\Omega = \dfrac{eB_0}{m_i c}$ is the ion gyrofrequency. In this case, n_{e_0}, n_{i_0} represent number densities of electron and ion, respectively, in the unperturbed state. In order to obtain the dispersion relation for low frequency waves, one can write the dependent variables as sum of equilibrium and perturbed parts. Writing, $n = 1 + \bar{n}$, $v_x = \bar{v}_x$, $v_z = \bar{v}_z$, $v_y = \bar{v}_y$ and $\phi = \bar{\phi}$, the equations (4.24)-(4.28) can be written as

$$\frac{\partial \bar{n}}{\partial t} + \frac{\partial \bar{v}_x}{\partial x} + \frac{\partial \bar{v}_z}{\partial z} = 0, \tag{4.29}$$

$$\frac{\partial \bar{v}_x}{\partial t} = -\frac{\partial \bar{\phi}}{\partial x} + \bar{v}_y, \tag{4.30}$$

$$\frac{\partial \bar{v}_y}{\partial t} = -\bar{v}_x, \tag{4.31}$$

$$\frac{\partial \bar{v}_z}{\partial t} = -\frac{\partial \bar{\phi}}{\partial z}, \tag{4.32}$$

$$\left(\frac{\partial^2}{\partial x^2} + \frac{\partial^2}{\partial z^2}\right)\bar{\phi} = \beta\left[\frac{q+1}{2}\bar{\phi} - \delta_1\bar{n}\right]. \tag{4.33}$$

One can presume that the perturbation is of the form $e^{i(k_x x + k_z z - \omega t)}$, where k_x (k_z) is the wave number in x (z) direction. Here, ω is the wave frequency ($\omega \ll \Omega$) for ion-acoustic wave obtained as

$$\omega = k_z\left[\frac{q+1}{2\delta_1} \cdot \frac{1}{\delta_1} + \left(1 + \frac{1}{\beta\delta_1}\right)k_x^2 + \left(1 + \frac{1}{\beta\delta_1}\right)k_z^2\right]^{-1/2}. \tag{4.34}$$

4.3.2 Derivation of the KP equation

To obtain the KP equation, one can employ RPT. The independent variables are stretched as follows:

$$X = \varepsilon^2 x, \tag{4.35}$$

$$\xi = \varepsilon(z - Vt), \tag{4.36}$$

$$\tau = \varepsilon^3 t, \tag{4.37}$$

where V denotes phase velocity of ion-acoustic wave and ε is a small parameter measuring the strength of the nonlinearity.

The dependent variables are expanded as

$$n = 1 + \varepsilon^2 n_1 + \varepsilon^4 n_2 + \cdots \tag{4.38}$$

$$v_x = \varepsilon^3 v_{x1} + \varepsilon^5 v_{x2} + \cdots \tag{4.39}$$

$$v_y = \varepsilon^3 v_{y1} + \varepsilon^5 v_{y2} + \cdots \tag{4.40}$$

$$v_z = \varepsilon^2 v_{z1} + \varepsilon^4 v_{z2} + \cdots \tag{4.41}$$

$$\phi = \varepsilon^2 \phi_1 + \varepsilon^4 \phi_2 + \cdots \tag{4.42}$$

Substituting the expansions given by equations (4.38)-(4.42) into the equations (4.29)-(4.32) and equating the coefficients of lowest power of ε, one gets the expression for phase velocity as

$$V^2 = \frac{2\delta_1}{1+q}. \tag{4.43}$$

Again, comparing higher order terms of ε, one gets

$$\frac{\partial n_1}{\partial \tau} - V\frac{\partial n_2}{\partial \xi} + \frac{\partial v_{x1}}{\partial X} + \frac{\partial v_{z2}}{\partial \xi} + \frac{\partial(n_1 v_{z1})}{\partial \xi} = 0, \tag{4.44}$$

$$\frac{\partial v_{z1}}{\partial \tau} - V\frac{\partial v_{z2}}{\partial \xi} + v_{z1}\frac{\partial v_{z1}}{\partial \xi} = -\frac{\partial \phi_2}{\partial \xi}, \tag{4.45}$$

$$\frac{\partial v_{x1}}{\partial \tau} - V\frac{\partial v_{x2}}{\partial \xi} + v_{z1}\frac{\partial v_{x1}}{\partial \xi} = -\frac{\partial \phi_2}{\partial X}, \tag{4.46}$$

$$\frac{\partial^2 \phi_1}{\partial \xi^2} = \beta\left[\frac{q+1}{2}\phi_2 + \frac{(q+1)(3-q)}{8}\phi_1^2 - \delta_1 n_2\right]. \tag{4.47}$$

From relations (4.44)-(4.47), the KP equation is obtained as

$$\frac{\partial}{\partial \xi}\left[\frac{\partial \phi_1}{\partial \tau} + A\phi_1\frac{\partial \phi_1}{\partial \xi} + B\frac{\partial^3 \phi_1}{\partial \xi^3}\right] + C\frac{\partial^2 \phi_1}{\partial X^2} = 0, \tag{4.48}$$

where $A = -\frac{(3-q)V^2-6}{4V}$, $B = \frac{V}{\beta(1+q)}$, and $C = \frac{V}{2}$.

4.3.3 *Formation of dynamical system and phase portraits*

One can obtain all travelling wave solutions of equation (4.48), considering the following transformation:

$$\phi_1(\xi, X, \tau) = \psi(\chi), \quad \chi = \alpha(l\xi + mX - U\tau), \tag{4.49}$$

where U is velocity of wave frame. Substituting equation (4.49) in the KP equation (4.48), integrating twice and disregarding integral constants, one can obtain ordinary differential equation as follows:

$$-a\psi + b\psi^2 + r\psi'' = 0, \tag{4.50}$$

where $a = lU - Cm^2$, $b = \frac{Al^2}{2}$, $r = B\alpha^2 l^4$ and ψ' is derivative of ψ with respect to χ.

One can have the following travelling wave system:

$$\begin{cases} \psi' = z, \\ z' = \frac{(a - b\psi)\psi}{r}, \end{cases} \tag{4.51}$$

which is a planar Hamiltonian system with Hamiltonian function:

$$H(\psi, z) = \frac{z^2}{2} - \frac{1}{6r}(3a - 2b\psi)\psi^2 = h, \text{say.} \tag{4.52}$$

As phase orbits given by the vector fields of equation (4.51) decide all travelling wave solutions of equation (4.48), one can examine bifurcations of phase profiles of equation (4.51) in the (ψ, z) phase plane as the parameters β, δ_1, q and U are varied. A solitary wave solution of equation (4.48) is analogous to a homoclinic orbit of equation (4.51). A periodic orbit of equation (4.51) corresponds to a periodic travelling wave solution of equation (4.48). The bifurcation theory of planar DSs plays an important role in our study (see [23]-[24]).

One can examine the bifurcations of phase profiles of equation (4.51). When $ab \neq 0$ and $r \neq 0$, there are two critical points at $E_0(\psi_0, 0)$ and $E_1(\psi_1, 0)$, where $\psi_0 = 0$ and $\psi_1 = \frac{a}{b}$. Let $M(\psi_i, 0)$ be the coefficient matrix of the linearised system of equation (4.51) at a critical point $E_i(\psi_i, 0)$. One can have

$$J = detM(\psi_i, 0) = -\frac{a}{r} + \frac{2b}{r}\psi_i. \tag{4.53}$$

By the theory of planar DSs, (see [23]-[24]), the critical point $E_i(\psi_i, 0)$ of the Hamiltonian system will be a saddle point if $J < 0$; the critical point $E_i(\psi_i, 0)$ of the Hamiltonian system will be a center if $J > 0$; when $J = 0$ and the Poincare index of the equilibrium point be 0, then the equilibrium point $E_i(\psi_i, 0)$ will be a cusp.

Considering $0 < l < 1$ and $\alpha \neq 0$, one can have the following cases.

Case 1: When $q > -1, \delta_1 > 0, \beta > 0, U < \frac{1-l^2}{l}\sqrt{\frac{\delta_1}{2(1+q)}}, 3(1+q) < \delta_1(3 - q)$, the system (4.51) possesses two critical points $E_0(\psi_0, 0)$ and $E_1(\psi_1, 0)$, where $\psi_0 = 0$ and $\psi_1 > 0$. Here, the critical points $E_0(\psi_0, 0)$ and $E_1(\psi_1, 0)$ are center and saddle point. It is seen that there is a homoclinic orbit through $E_1(\psi_1, 0)$ surrounding the center $E_0(\psi_0, 0)$(see Figure 4.7).

Case 2: When $q > -1, \delta_1 > 0, \beta > 0, U > \frac{1-l^2}{l}\sqrt{\frac{\delta_1}{2(1+q)}}, 3(1+q) > \delta_1(3 - q)$, the system (4.51) has two equilibrium points $E_0(\psi_0, 0)$ and $E_1(\psi_1, 0)$, where $\psi_0 = 0$ and $\psi_1 > 0$. Here, $E_0(\psi_0, 0)$ is a saddle point and $E_1(\psi_1, 0)$ is a center. It is seen that there is a homoclinic orbit through $E_0(\psi_0, 0)$ enclosing the center $E_1(\psi_1, 0)$ (see Figure 4.8).

Case 3: When $q > -1, \delta_1 > 0, \beta > 0, U > \frac{1-l^2}{l}\sqrt{\frac{\delta_1}{2(1+q)}}, 3(1+q) < \delta_1(3 - q)$, the system (4.51) possesses two critical points $E_0(\psi_0, 0)$ and $E_1(\psi_1, 0)$, where $\psi_0 = 0$ and $\psi_1 < 0$. Here, $E_0(\psi_0, 0)$ and $E_1(\psi_1, 0)$ are saddle point and center, respectively. In this case, one can observe that there is a homoclinic orbit through $E_0(\psi_0, 0)$ surrounding the center $E_1(\psi_1, 0)$ (see Figure 4.9).

Case 4: When $q > -1, \delta_1 > 0, \beta > 0, U < \frac{1-l^2}{l}\sqrt{\frac{\delta_1}{2(1+q)}}, 3(1+q) > \delta_1(3 - q)$, the system (4.51) possesses two equilibrium points $E_0(\psi_0, 0)$ and $E_1(\psi_1, 0)$, where $\psi_0 = 0$ and $\psi_1 < 0$. Here, $E_0(\psi_0, 0)$ and $E_1(\psi_1, 0)$ are center and saddle point, respectively. One can see the existence of a homoclinic orbit through $E_1(\psi_1, 0)$ surrounding the center $E_0(\psi_0, 0)$ (see Figure 4.10).

Using the analysis above, different phase profiles of equation (4.51) can be observed depending on some particular parameter values, given in Figures 4.7-4.10.

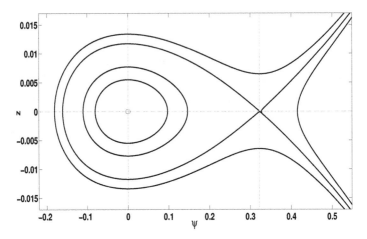

Figure 4.7: Phase portrait of equation (4.51) for $\alpha = 5, l = 0.34, \beta = 0.9$, $\delta_1 = 0.41, q = -0.9$ and $U = 3.6$.

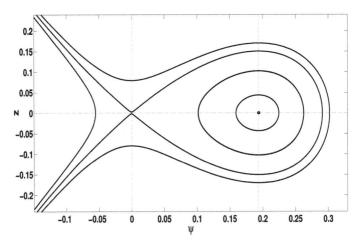

Figure 4.8: Phase portrait of equation (4.51) for $\alpha = 2, l = 0.3, \beta = 0.6$, $\delta_1 = 0.2, q = 0.8$ and $U = 0.8$.

4.3.4   Wave solutions

Two types of exact travelling wave solutions of equation (4.48) can be obtained by employing travelling wave transformation to equation (4.51) and the Hamiltonian (4.52) with $h = 0$. These solutions are solitary wave solution and periodic travelling wave solution.

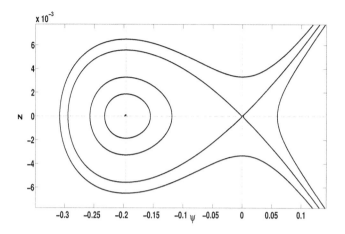

Figure 4.9: Phase portrait of equation (4.51) for $\alpha = 5, l = 0.34, \beta = 0.9$, $\delta_1 = 0.41, q = -0.9$ and $U = 3.8$.

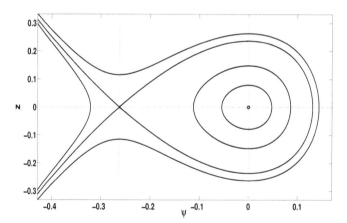

Figure 4.10: Phase portrait of equation (4.51) for $\alpha = 2, l = 0.3, \beta = 0.6$, $\delta_1 = 0.2, q = 0.8$ and $U = 0.6$.

(i) The snoidal wave solution for the equation (4.48) analogous to a periodic trajectory of the phase portrait exhibited in Figure 4.7 is obtained as

$$\psi = \frac{-R_2 + R_1 \left(\frac{R_2 - R_3}{R_1 - R_3} sn^2 \left(\frac{K_1}{g} \chi, k \right) \right)}{-1 + \left(\frac{R_2 - R_3}{R_1 - R_3} sn^2 \left(\frac{K_1}{g} \chi, k \right) \right)}, \tag{4.54}$$

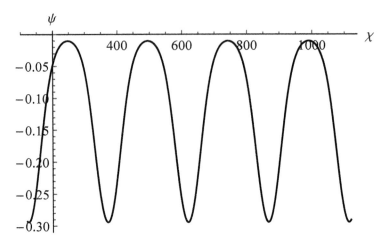

Figure 4.11: Periodic wave profile of the KP equation (4.48) for $\alpha = 5$, $l = 0.34$, $\beta = 0.9, \delta_1 = 0.41$, $U = 0.38$ and $q = -0.9$.

where $K_1 = \sqrt{\frac{-2b}{3r}}$. The above solution (4.54) exists for $b < 0$ and $R_1 > R_2 > \psi \geq R_3$ [25], where R_1, R_2 and R_3 are roots of $h_i + \frac{1}{6r}(3a - 2b\psi)\psi^2 = 0$ with $h_i = H(\psi_i, z_i)$ at any point (ψ_i, z_i) on the periodic trajectory of the phase portrait displayed in Figure 4.7. Here, *sn* denotes Jacobian elliptic function, $g = \frac{2}{\sqrt{R1 - R_3}}$ and $k = \sqrt{\frac{R_2 - R_3}{R_1 - R_3}}$.

(ii) When the conditions in Case 2 are satisfied (see Figures (4.8) and (4.12)), the system (4.51) has a smooth solitary wave solution of the form:

$$\psi = \frac{3a}{2b} \operatorname{sech}^2\left(\frac{1}{2}\sqrt{\frac{a}{r}}\chi\right). \tag{4.55}$$

Graphs of the exact solutions (4.54) and (4.55) of the system (4.51) depending on some particular values of the parameters are shown in Figures 4.11-4.12, respectively.

In Figure 4.11, for the periodic travelling wave solution of equation (4.48), ψ is plotted against χ for $\alpha = 5$, $l = 0.34$, $\beta = 0.9, \delta_1 = 0.41$, $U = 0.38$ and $q = -0.9$ (Here, parameters are chosen so that they satisfy the conditions of Case 1). In case of Figure 4.12, for the solitary wave solution of equation (4.48), ψ is plotted against χ for $\alpha = 2, l = 0.3, \beta = 0.6, \delta_1 = 0.2, U = 0.8$ and $q = 0.8$ (Here, parameters are chosen so that they satisfy the conditions of Case 2).

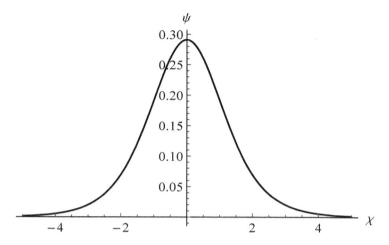

Figure 4.12: Solitary wave profile of the KP equation (4.48) for $\alpha = 2$, $l = 0.3, \beta = 0.6, \delta_1 = 0.2, U = 0.8$ and $q = 0.8$.

4.4 Bifurcation of dust-acoustic waves with small amplitude

The study of dust-acoustic waves (DAWs) in a variety of dusty plasmas has gained immense popularity due to its wide applications in laboratory and space plasmas. One can state that dusty plasmas are strongly coupled when the condition $\Gamma > 1$ is satisfied and weakly coupled when the condition $\Gamma < 1$ is satisfied. Here, coupling parameter is denoted by Γ [26]. Dust particles are caused to shift from gaseous state to crystal state by the strong coupling [27, 28]. Dusty plasmas having strongly coupled composition was experimentally observed by many researchers [10, 29, 30, 31]. In 1997, DAWs in strongly coupled dusty plasmas (SCDPs) was reported by Rosenberg and Kalman [32]. Dispersion of DAWs was investigated by Pieper and Goree [33] in strongly correlated plasmas. Effects of collision and magnetic field in SCDPs were studied by Kaw [34]. Investigation of nonlinear waves in SCDPs through experimental observations was carried out by some researchers [35, 36]. Breaking of DAWs in SCDPs was examined by Shukla and Lin [37]. Impacts of dust and trapped ion temperatures on DAWs were inspected by Alinejad [38]. Alfvenic turbulence was investigated by Misra and Banerjee [39] in a magnetized dusty plasmas. Effects of heavy ions and nonextensivity were reported by Ema et al. [40] in strongly coupled plasmas. Humped DASWs were studed by El-Borie and Atteya [41] in strongly coupled cryogenic dusty plasmas. Very recently, variation of DA freak waves related to nonplanar geometry was reported by Almutalk et al. [42].

4.4.1 Basic equations

Propagation of DAWs in SCDP comprising of Maxwellian electrons, strongly coupled dusts [43] and ions is described by the following basic equations [34]:

$$\frac{\partial n_d}{\partial t} + \frac{\partial}{\partial x}(n_d u_d) = 0, \tag{4.56}$$

$$\left(1 + \tau_m \frac{\partial}{\partial t}\right)\left[n_d\left(D_t u_d + v_{dn} u_d - \frac{\partial \phi}{\partial x}\right)\right] = \eta \frac{\partial^2 u_d}{\partial x^2}, \tag{4.57}$$

$$\frac{\partial^2 \phi}{\partial x^2} = n_d + \alpha_e e^{\phi} - \alpha_i e^{-\sigma \phi}. \tag{4.58}$$

where n_d, n_i and n_e are, respectively, number densities of dusts, ions and electrons, and are, respectively, normalized by n_{d0}, n_{i0} and n_{e0}. Here, u_d and ϕ are, respectively, velocity of dusts and electrostatic potential, and are, respectively, normalized by $C_d = \sqrt{\frac{Z_d T_e}{m_d}}$ and $\frac{T_e}{e}$, where T_e denotes temperature of electrons and e denotes measure of electron charge. Here, $D_t \equiv \frac{\partial}{\partial t} + u_d \frac{\partial}{\partial x}$, $\sigma = \frac{T_i}{T_e}$, $\alpha_e = \frac{n_{e0}}{Z_d n_{d0}}$ and $\alpha_i = \frac{n_{i0}}{Z_d n_{d0}}$, where T_i is temperature of ions, Z_d is the number of charged particles residing on the surface of dusts. Here, v_{dn} stands for collisional frequency of dusts and neutral particles normalized by τ_d^{-1}. In this case, η is the coefficient of longitudinal viscosity which is given by $(\frac{\tau_d}{m_d n_{d0} \lambda_{Dd}^2})[\eta_t + (4/3)\zeta_t]$, where ζ_t and η_t are transport coefficients of bulk and shear viscosities, respectively. Here, time variable is denoted by t and is normalized by $\tau_d = (\frac{m_d}{4\pi n_{d0} Z_d^2 e^2})^{\frac{1}{2}}$, and x is space variable which is normalized by $\lambda_{Dd} = (\frac{T_d}{4\pi Z_d n_{d0} e^2})^{\frac{1}{2}}$. Here, τ_m denotes the relaxation time of viscoelasticity and is normalized by τ_d, where expression for τ_m is [44]

$$\tau_m = \eta \frac{T_e}{T_d}\left[1 - \mu_d + \frac{4}{15}u(\Gamma)\right]^{-1}, \tag{4.59}$$

with

$$\mu_d = 1 + \frac{1}{3}u(\Gamma) + \frac{\Gamma}{9}\frac{\partial u(\Gamma)}{\partial \Gamma}, \tag{4.60}$$

being the compressive factor [44]. The compressive factor [45] is generated by considering macroscopic illustration of the response of the system. Here, $u(\Gamma)$ denotes strength of surplus inner energy given by $u(\Gamma) \simeq -(\sqrt{3}/2)\Gamma^{3/2}$ for $\Gamma < 1$ (weakly coupled plasmas) and for $1 < \Gamma < 100$ (strongly coupled plasmas), $u(\Gamma)$ is given by Slattery et al. [46]

$$u(\Gamma) \simeq -0.89\Gamma + 0.95\Gamma^{1/4} + 0.19\Gamma^{-1/4} - 0.81. \tag{4.61}$$

However, dependency of η on Γ was presented in the works [31, 47].

4.4.2 Derivation of the Burgers equation

The Burgers equation is obtained employing RPT with following stretching [49, 48, 43, 50] based on dispersion relation between wave frequency and wave number

$$\xi = \varepsilon(x - Vt), \quad \tau = \varepsilon^2 t, \tag{4.62}$$

where ε stands for measure of weakness of dispersion and V denotes phase velocity of the wave. One can expand n_d, u_d and ϕ as follows

$$\begin{cases} n_d = 1 + \varepsilon n_{d1} + \varepsilon^2 n_{d2} + \cdots \\ u_d = 0 + \varepsilon u_{d1} + \varepsilon^2 u_{d2} + \cdots \\ \phi = 0 + \varepsilon \phi_1 + \varepsilon^2 \phi_2 + \cdots . \end{cases} \tag{4.63}$$

The initial condition to obtain wave solutions are given by

$$n_d \to 1, \ u_d \to 0, \ \phi \to 0 \ \text{ when } \ |\xi| \to \infty. \tag{4.64}$$

Using equations (4.62)-(4.63) into governing equations (4.56)-(4.57), one can obtain the following relations from coefficients of ε^2

$$\begin{cases} n_{d1} = -\frac{\phi_1}{V^2}, \\ u_{d1} = -\frac{\phi_1}{V}, \\ \frac{1}{V^2} = \alpha_e + \sigma \alpha_i. \end{cases} \tag{4.65}$$

From equation (4.65), one can observe that phase velocity V depends on α_e, α_i and temperature ratio σ.

Comparing the coefficients of ε^3 and removing n_{d2}, u_{d2} and ϕ_2, one can obtain the Burgers equation as

$$\frac{\partial \phi}{\partial \tau} + A\phi_1 \frac{\partial \phi_1}{\partial \xi} = B \frac{\partial^2 \phi_1}{\partial \xi^2}, \tag{4.66}$$

where $A = \dfrac{V^3}{2b}\left[\dfrac{1}{V^4}(2\tau_m V_{dn} - 3) - a\right]$, $B = \dfrac{\eta}{2b}$, $a = (\alpha_e - \sigma^2 \alpha_i)$ and $b = \left(1 + \dfrac{\tau_m V_{dn}}{2}\right)$.

4.4.3 Formation of dynamical system and phase portraits

For bifurcation analysis [23, 24], one can transform the Burgers equation (4.66) into a DS by employing the transformation

$$\phi_1(\xi, \tau) = \psi(\chi), \quad \chi = \xi - \lambda \tau, \tag{4.67}$$

where λ denotes traveling wave velocity. Substituting equation (4.67) into equation (4.66) and integrating with respect to χ under the condition (4.64), the Burgers equation (4.66) is transformed into DS as

$$\begin{cases} \dfrac{d\psi}{d\chi} = z, \\ \dfrac{dz}{d\chi} = \dfrac{A^2}{2B^2}\psi\left(\psi - \dfrac{\lambda}{A}\right)\left(\psi - \dfrac{2\lambda}{A}\right). \end{cases} \tag{4.68}$$

It can be observed that equation (4.68) has three fixed points $E_0 = (0,0)$, $E_1 = \left(\frac{\lambda}{A}, 0\right)$ and $E_2 = \left(\frac{2\lambda}{A}, 0\right)$. At E_0 and E_2, determinants of Jacobian matrix (M) are less than 0 that imply existence of a pair of saddle points. While determinant of Jacobian matrix (M) is greater than 0 at fixed point E_1, and this implies that E_1 is a center. Now, using the phase plane analysis of DS, one can present a phase plot in Figure 4.13 for equation (4.68) of the Burgers equation (4.66) with parameters α_e, α_i, η, V_{dn}, Γ, T_e, T_i, T_d and λ.

Phase profile of equation (4.68), displayed in Figure 4.13, it is perceived that saddle points at E_0 and E_2 are connected by heteroclinic orbits enclosing center at E_1. Heteroclinic orbits correspond to shock waves, namely, kink and anti-kink waves, whereas, the periodic orbit corresponds to periodic wave. Hence, it is evident that the Burgers equation in the considered SCDP exhibits kink, anti-kink and periodic solutions with parameters α_e, α_i, η, V_{dn}, Γ, T_e, T_i, T_d and λ.

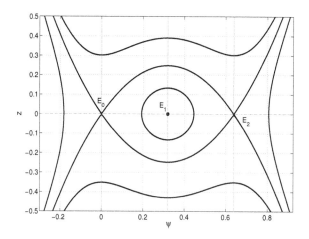

Figure 4.13: Phase portrait of system (4.68) for $\alpha_e = 0.3$, $\alpha_i = 0.6$, $\eta = 0.7$, $v_{dn} = 0.7$, $T_e = 0.4$, $T_i = 0.3$, $T_d = 0.4$, $\lambda = 0.2$, $\Gamma = 1.2$ and $v = 0.1$.

4.4.4 *Wave solutions*

To achieve analytical dust-acoustic kink waves (DAKWs) and dust-acoustic anti-kink waves (DAAKWs) solutions, one can consider

$$\zeta = c(\xi - v\tau), \tag{4.69}$$

as new variable, where v is velocity of the wave and $c > 0$. Applying ζ and solving with the *tanh* method [51] on the Burgers equation (4.66), one can obtain the analytical shock wave solutions, i.e., DAKW and DAAKW solutions as

$$\psi(\xi, \tau) = \pm \frac{v}{A}\left[1 - tanh\left\{\frac{v}{2B}(\xi - v\tau)\right\}\right]. \tag{4.70}$$

Here, the amplitude of the DAW is $\frac{v}{A}$. The DAKW and DAAKW solutions are presented in Figures 4.14-4.17 by varying parameters Γ and η, keeping $\tau = 1$ with other parameters fixed. The DAKW and DAAKW solutions in SCDP using the *tanh* function are reported in the work [43]. Therefore, in this work, both DAKW and DAAKW solutions of the Burgers equation (4.66) are supported by the work [43].

Figures 4.14 and 4.15 depict the influence of Γ on DAKW and DAAKW solutions, respectively, for $\alpha_e = 0.3$, $\alpha_i = 0.6$, $\eta = 0.7$, $v_{dn} = 0.7$, $T_e = 0.4$, $T_i = 0.3$, $T_d = 0.4$, $\lambda = 0.2$ and $v = 0.1$. As values of coupling parameter

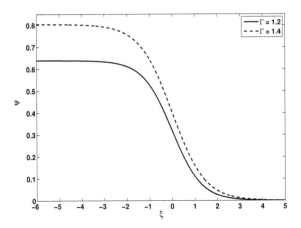

Figure 4.14: Kink wave solution of equation (4.66) by varying Γ.

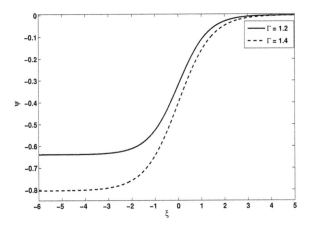

Figure 4.15: Anti-Kink wave solution of equation (4.66) by varying Γ.

Γ increases in strongly coupled region ($\Gamma > 1$), then the amplitudes of both DAKWs and DAAKWs rise while smoothness declines.

Figures 4.16 and 4.17 show the influence of η on DAKWs and DAAKWs, respectively, for $\alpha_e = 0.3$, $\alpha_i = 0.6$, $\Gamma = 1.2$, $v_{dn} = 0.7$, $T_e = 0.4$, $T_i = 0.3$, $T_d = 0.4$, $\lambda = 0.2$ and $v = 0.1$. When there is growth in longitudinal viscosity coefficient (η), amplitudes of both DAKWs and DAAKWs fall while smoothness increases.

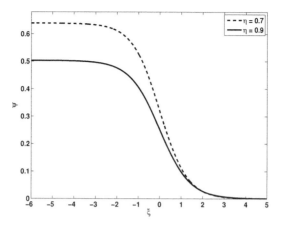

Figure 4.16: Kink wave solution of equation (4.66) by varying Γ.

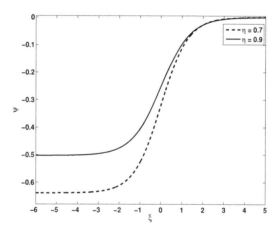

Figure 4.17: Anti-Kink wave solution of equation (4.66) by varying η.

In Figure 4.13, a family of nonlinear periodic trajectories about the fixed point E_1 is presented that corresponds to a family of periodic solutions of the Burgers equation (4.66). Thus, to acquire the dust-acoustic periodic wave (DAPW) solution, one can find the Hamiltonian function of the dynamical system (4.68) as

$$H(\psi, z) = \frac{z^2}{2} - \frac{A^2}{2B^2}\left(\frac{1}{4}\psi^4 - \frac{\lambda}{A}\psi^3 + \frac{\lambda^2}{A^2}\psi^2\right) = h, \qquad (4.71)$$

and one can obtain

$$\frac{d\psi}{d\chi} = \frac{A}{2B}\sqrt{(p-\psi)(\psi-q)(\psi-r)(\psi-s)}. \tag{4.72}$$

Here p, q, r and s are roots of $h_i + \frac{A^2}{2B^2}\left(\frac{1}{4}\psi^4 - \frac{\lambda}{A}\psi^3 + \frac{\lambda^2}{A^2}\psi^2\right) = 0$, with $h_i = H(\psi_i, z_i)$ for a point (ψ_i, z_i) on the periodic trajectory around E_1 of Figure 4.13. The DAPW solution of the Burgers equation can be acquired by using (4.72) as

$$\psi = \frac{p+s\left\{\dfrac{p-q}{q-s}sn^2\left(\dfrac{A}{2Bg}\chi,k\right)\right\}}{1+\dfrac{p-q}{q-s}sn^2\left(\dfrac{A}{2Bg}\chi,k\right)}, \tag{4.73}$$

where $g = \dfrac{2}{\sqrt{(p-r)(q-s)}}$ and $k = \sqrt{\dfrac{(p-q)(r-s)}{(p-r)(q-s)}}$. Here sn is the Jacobi elliptic function [25]. The bistable periodic solitary waves for DAWs in dusty plasmas was reported in the work [52]. Thus, using the Jacobi elliptic function one can obtain DAPW solution (4.73) of the Burgers equation (4.66).

Figures 4.18-4.19 display changes on the DAPW solution (4.73) of the Burgers equation (4.66) by varying values of parameters Γ and η, respectively while keeping $\alpha_e = 0.3$, $\alpha_i = 0.6$, $v_{dn} = 0.7$, $T_e = 0.4$, $T_i = 0.3$, $T_d = 0.4$ and $\lambda = 0.2$ fixed. Figure 4.18 shows rise in amplitude and width of DAPWs as values of $\Gamma = 1.1$ (dotted), 1.15 (continuous) increase.

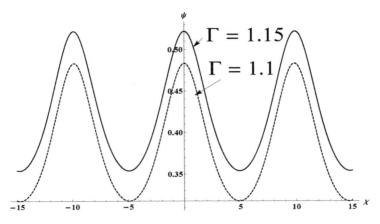

Figure 4.18: Periodic wave solution of equation (4.66) by varying Γ.

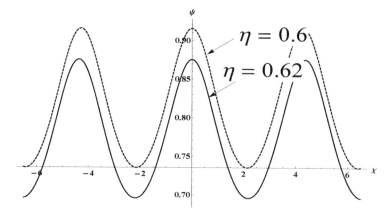

Figure 4.19: Periodic wave solution of equation (4.66) by varying η.

Figure 4.19 shows that with rise in values of longitudinal viscosity effect $\eta = 0.6$ (dotted), 0.62 (continuous), one can observe decrease in amplitude and width of DAPWs.

4.5 Bifurcation of electron-acoustic waves with small amplitude

Over the past few years, the investigation of electrostatic waves in quantum plasma has flourished and it has become an exciting research subject in the field of plasma physics. An electrostatic wave, known as electron-acoustic waves (EAWs), is a wave with a relatively higher frequency than the ion plasma. EAWs can be observed in laboratory for a plasma that consists of cold and hot electrons [53, 54] and in electron-ion plasma composed of ions with higher temperature than electrons [55]. The transmission of EAWs performs an important part in laboratory and space plasmas. Many researchers studied linear [56, 57, 58] and nonlinear properties [59, 60, 61, 62, 63] of EAWs in unmagnetized plasma. Electron-acoustic solitary wave (EASW) was investigated in unmagnetized plasma having ions with finite temperature [60]. In addition to that, numerous studies have been reported on investigation of EAW features in plasmas with magnetic effects [64, 65, 66]. The fusion devices and auroral ionosphere [67] contain two temperature electrons (one hot and one cold) and this leads to the existence of EAW in such plasmas. EAWs are capable of describing the electrostatic component of the broadband electrostatic noise (BEN) that are found in magnetosphere [68] and geomagnetic tail [69].

4.5.1 Basic equations

Consider a quantum plasma that contains cold electrons, hot electrons and ions and has no magnetic and collisional effects. The motion of cold electrons is considered. The phase speed of EAWs satisfies $v_{Fc} \ll \omega/k \ll v_{Fh}$, where v_{Fc} and v_{Fh} are Fermi velocities of cold and hot electrons, respectively. The normalized basic equations [70] are:

$$\frac{\partial n}{\partial t} + \frac{\partial (nu)}{\partial x} = 0, \tag{4.74}$$

$$\frac{\partial u}{\partial t} + u\frac{\partial u}{\partial x} = \alpha\frac{\partial \phi}{\partial x}, \tag{4.75}$$

$$\frac{\partial^2 \phi}{\partial x^2} = n_h + \frac{1}{\alpha}n - \left(1 + \frac{1}{\alpha}\right), \tag{4.76}$$

$$\phi = -\frac{1}{2} + \frac{n_h^2}{2} + \frac{H^2}{6\sqrt{n_h}}\frac{\partial^2 \sqrt{n_h}}{\partial x^2}, \tag{4.77}$$

where n_0 normalizes the cold electron number density n, n_{h0} normalizes hot electron number density n_h, $\sqrt{2k_B T_{Fh}/\alpha m_e}$ normalizes the cold electron velocity u where, $\alpha = n_{h0}/n_0 > 1$, m_e is the electron mass, T_{Fh} is the Fermi temperature of hot electron given by the relation $m_e v_{Fh}^2/2 = k_B T_{Fh}$, e is the electron charge, k_B is the Boltzman constant. Here, $2k_B T_{Fh}/e$ normalizes the electrostatic wave potential ϕ. The nondimensional quantum parameter $H = \hbar\omega_{ph}/2k_B T_{Fh}$, $\omega_{ph} = \sqrt{4\pi n_{h0}e^2/m_e}$ is the hot electron plasma frequency, $\hbar$ is Planck's constant. Fermi wave length of hot electron $\lambda_{Fh} = \sqrt{2k_B T_{Fh}/4\pi n_{h0}e^2}$ normalizes space variable x and $\omega_{pc}^{-1} = \sqrt{m_e/4\pi n_0 e^2}$ normalizes time. The equilibrium condition holds $n_0 + n_{h0} = n_{i0}$.

4.5.2 Derivation of the KdV equation

The reductive perturbation technique (RPT) is applied to derive the KdV equation that defines the features of finite amplitude EAWs in this quantum plasma. According to RPT, the following stretched coordinates are used

$$\xi = \varepsilon^{\frac{1}{2}}(x - vt), \tag{4.78}$$

$$\tau = \varepsilon^{\frac{3}{2}}t, \tag{4.79}$$

where v represents the phase velocity of EAW and ε is a small parameter that defines the strength of the nonlinearity. Expanding the dependent quantities, one can have

$$n = 1 + \varepsilon n_1 + \varepsilon^2 n_2 + \dots \tag{4.80}$$

$$n_h = 1 + \varepsilon n_h^{(1)} + \varepsilon^2 n_h^{(2)} + \dots \tag{4.81}$$

$$u = \varepsilon u_1 + \varepsilon^2 u_2 + \dots \tag{4.82}$$

$$\phi = \varepsilon \phi_1 + \varepsilon^2 \phi_2 + \dots \tag{4.83}$$

Using equations (4.78)-(4.83) into the system of equations (4.74)-(4.77) and comparing the terms of lowest power of ε, one can get

$$n_1 = \frac{1}{v} u_1, \tag{4.84}$$

$$u_1 = -\frac{\alpha}{v} \phi_1, \tag{4.85}$$

$$n_h^{(1)} = -\frac{1}{\alpha} n_1, \tag{4.86}$$

$$\phi_1 = n_h^{(1)}. \tag{4.87}$$

Comparing the terms of next power of ε, one can get

$$\frac{\partial n_1}{\partial \tau} - v \frac{\partial n_2}{\partial \xi} + \frac{\partial}{\partial \xi}(n_1 u_1) + \frac{\partial u_2}{\partial \xi} = 0, \tag{4.88}$$

$$\frac{\partial u_1}{\partial \tau} - v \frac{\partial u_2}{\partial \xi} + u_1 \frac{\partial u_1}{\partial \xi} = \alpha \frac{\partial \phi_2}{\partial \xi}, \tag{4.89}$$

$$\frac{\partial^2 \phi_1}{\partial \xi^2} = n_h^{(2)} + \frac{1}{\alpha} n_2, \tag{4.90}$$

$$n_h^{(2)} = \phi_2 - \frac{H^2}{12} \frac{\partial^2 n_h^{(1)}}{\partial \xi^2} - \frac{(n_h^{(1)})^2}{2}. \tag{4.91}$$

From the relations (4.84)-(4.87), one can get

$$v^2 = 1. \tag{4.92}$$

From the relations (4.87)-(4.92), one can obtain the KdV equation as

$$\frac{\partial \phi_1}{\partial \tau} + \frac{v}{2}(1 - 3\alpha)\phi_1 \frac{\partial \phi_1}{\partial \xi} + \frac{v}{2}\left(1 + \frac{H^2}{12}\right)\frac{\partial^3 \phi_1}{\partial \xi^3} = 0. \tag{4.93}$$

4.5.3 Formation of dynamical system and phase portraits

One can form the dynamical system of the KdV equation (4.93) by using new variable χ as

$$\chi = (\xi - c\tau), \tag{4.94}$$

where c stands for velocity of the traveling wave. Setting $\psi(\chi) = \phi_1(\xi, \tau)$ and putting into the KdV equation (4.93), one can have

$$-c\frac{d\psi}{d\chi} + \frac{v}{2}(1 - 3\alpha)\psi\frac{d\psi}{d\chi} + \frac{v}{2}\left(1 + \frac{H^2}{12}\right)\frac{d^3\psi}{d\chi^3} = 0. \tag{4.95}$$

Integrating equation (4.95) w.r.t χ and neglecting integral constant, one can have

$$-c\psi + \frac{v}{4}(1 - 3\alpha)\psi^2 + \frac{v}{2}\left(1 + \frac{H^2}{12}\right)\frac{d^2\psi}{d\chi^2} = 0. \tag{4.96}$$

Then equation (4.96) is framed into the following dynamical system:

$$\begin{cases} \frac{d\psi}{d\chi} = z, \\ \frac{dz}{d\chi} = \frac{1}{v\left(1 + \frac{H^2}{12}\right)}\left(2c - \frac{v}{2}(1 - 3\alpha)\psi\right)\psi. \end{cases} \tag{4.97}$$

The equation (4.97) is a planar dynamical system with Hamiltonian function:

$$H_m(\psi, z) = \frac{z^2}{2} - \frac{1}{3v\left(1 + \frac{H^2}{12}\right)}(3c - \frac{v}{2}(1 - 3\alpha)\psi)\psi^2 = h, \text{say}. \tag{4.98}$$

The planar dynamical system equation (4.97) is influenced by parameters α, v, H and c. The phase orbits defined in the vector fields of equation (4.97) describe all traveling wave solutions of the KdV equation (4.93). For different values of α, v, H and c, bifurcations of phase portraits of equation (4.97) are examined in the (ψ, z) phase plane. For this analysis, one may consider a physical system for which only bounded traveling wave solutions are significant. A solitary wave solution of equation (4.93) corresponds to a homoclinic orbit of equation (4.97). A periodic orbit of equation (4.97) corresponds to a periodic traveling wave solution of equation (4.93). The bifurcation theory and methods of planar dynamical systems play a significant role in this work [23]-[24].

One can study the bifurcations of phase portraits of the dynamical system (4.97). For $H^2 \neq 4$ and $\alpha > 1$, two fixed points are obtained, $E_0(\psi_0, 0)$ and

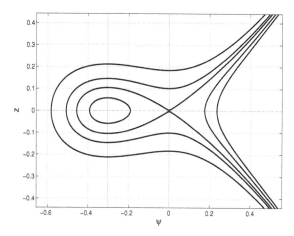

Figure 4.20: Phase portrait of equation (4.97) for $\alpha = 1.2, v = 1, H = 1.001$ and $c = 0.808$.

$E_1(\psi_1, 0)$, where $\psi_0 = 0$ and $\psi_1 = \frac{4c}{v(1-3\alpha)}$. Let $M(\psi_i, 0)$ be the coefficient matrix of the linearized system of the dynamical system (4.97) at a fixed point $E_i(\psi_i, 0)$. Then one can have

$$J = detM(\psi_i, 0) = -\frac{2c}{v\left(1 + \frac{H^2}{12}\right)} + \frac{1 - 3\alpha}{\left(1 + \frac{H^2}{12}\right)}\psi_i. \qquad (4.99)$$

Analyzing through the theory of planar dynamical systems [23]-[24], fixed point $E_i(\psi_i, 0)$ is a saddle point when $J < 0$, and the fixed point $E_i(\psi_i, 0)$ is a center when $J > 0$.

For $v = 1, c > 0, \alpha > 1$ and $H^2 < 4$, the dynamical system equation (4.97) contains two equilibrium points at $E_0(\psi_0, 0)$ and $E_1(\psi_1, 0)$, where $\psi_0 = 0$ and $\psi_1 < 0$. Here $E_0(\psi_0, 0)$ is a saddle point and $E_1(\psi_1, 0)$ is a center. There is a homoclinic orbit about an equilibrium point $E_0(\psi_0, 0)$ enclosing the center $E_1(\psi_1, 0)$ (see Figure 4.20).

4.5.4 Wave solutions

Through the theory of planar dynamical system, equation (4.97) and equation (4.98) correspond to exact traveling wave solutions of equation (4.93), namely, the solitary wave solution and periodic traveling wave solution for different parameter values.

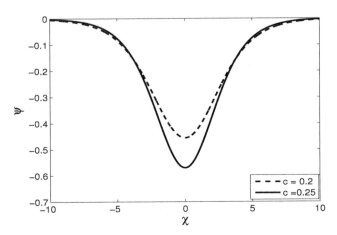

Figure 4.21: Graph of the solitary wave solution of equation (4.93) for $\alpha = 1.302, v = 1, H = 2.81$.

If $v = 1, c > 0, \alpha > 1$ and $H^2 < 4$, (see Figures 4.21 and 4.22), the system equation (4.93) has the solitary wave solution given by

$$\phi_1 = \frac{6c}{v(1 - 3\alpha)} sech^2(\frac{1}{2}\sqrt{\frac{2c}{v(1 + \frac{H^2}{12})}}\chi). \qquad (4.100)$$

By using symbolic computations, one can obtain two graphs of the solitary wave solution and periodic traveling wave solution of equation (4.93) depending on some particular values of the system parameters.

Figure 4.21 presents a graph for the solitary wave solution (4.100) in which ψ vs. χ is plotted for $\alpha = 1.302, v = 1$ and $H = 2.81$ with different values of c. Figure 4.22 presents a graph for the solitary wave solution (4.100) in which ψ vs. χ is plotted for $\alpha = 1.2, v = 1$ and $c = 0.808$ with different values of H. The parameter c has a remarkable influence on the amplitude and width of the solitary wave solution but the parameter H has no influence on its amplitude. Thus, the parameters c and H have significant influence on the solitary wave solution.

Figure 4.23 presents a graph for the periodic traveling wave solution in which ψ vs. χ is plotted for $\alpha = 1.302, v = 1$ and $H = 2.81$ with different values of c. Figure 4.24 presents a graph for the periodic traveling wave solution in which ψ vs. χ is plotted for $\alpha = 1.302, v = 1$ and $c = 1.2594$ with different

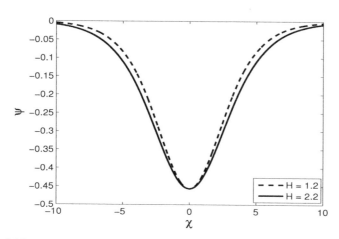

Figure 4.22: Graph of the solitary wave solution of equation (4.93) for $\alpha = 1.2, v = 1$ and $c = 0.808$.

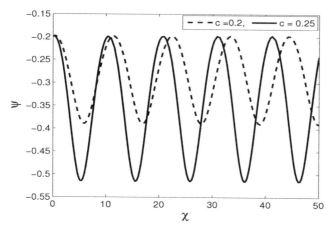

Figure 4.23: Graph of the periodic traveling wave solution of equation (4.93) for $\alpha = 1.302, v = 1$ and $H = 2.81$.

values of H. Amplitude of the periodic wave solution is influenced by c and is not affected by H. The width of the periodic traveling wave solution is influenced by both c and H.

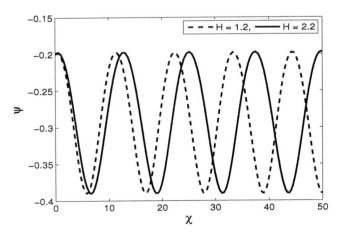

Figure 4.24: Graph of the periodic traveling wave solution of equation (4.93) for $\alpha = 1.302, v = 1$ and $c = 1.2594$.

References

[1] R. K. Dodd, J. C. Eilbeck, J. D. Gibbon and H. C. Morries, Solitons and Nonlinear Waves Equations, Academic Press Inc. (1982).

[2] A. A. Mamun, Astrophysics and Space Science, 268: 443–454 (1999).

[3] T. K. Maji, M. K. Ghorui, A. Saha and P. Chatterjee, Brazilian Journal of Physics, 47(3): 295–301 (2017).

[4] F. Verheest, C. P. Olivier and W. A. Hereman, J. Plasma Phys., 18: 905820208 (2018).

[5] W. D. Jones, A. Lee, S. M. Gleman and H. J. Doucet, Phys. Rev. Lett., 35: 1349 (1975).

[6] S. H. Strogatz, Nonlinear Dynamics and Chaos, Westview Press (USA) (2007).

[7] A. Saha and J. Tamang, Advances in Space Research, 63: 1596–1606 (2019).

[8] C. K. Goertz, Rev. Geophys., 27: 271, doi:10.1029/RG027i002p00271 (1989).

[9] T. G. Northrop and T.J. Bringham, Planet Space Science, 38: 319 (1990).

[10] J. H. Chu, J. B. Du and I. Lin, J. Physics D, 27: 296 (1994).

[11] J. H. Chu, J. B. Du and I. Lin, Phys. Rev. Lett., 72: 4009 (1994).

[12] H. Thomas, G. E. Morfill and V. Dammel, Phys. Rev. Lett., 73: 652 (1994).

[13] Y. Hayashi and K. Tachibana, Jpn. J. Appl. Phys. Part 2, 33: L804 (1994).

[14] P. K. Shukla and V. P. Slin, Phys. Scr., 45: 508 (1992).

[15] R. L. Merlino, A. Barkan, C. Thompson and N. D'Angelo, Phys. Plasmas, 5: 1607 (1998).

[16] N. N. Rao, P. K. Shukla and M. Y. Yu, Planet Space Sci., 38: 543 (1990).

[17] A. A. Mamun, Astrophys. Space Sci., 268: 443 (1999).

[18] P. K. Shukla, M. Y. Yu and R. Bharuthram, J. Geophys. Res., 96: 21343 (1992).

[19] F. Melandso, Phys. Plasmas, 3: 3890 (1996).

[20] P. K. Shukla and R. K. Varma, Phys. Fluids B, 5: 236 (1993).

[21] A. Barkan, R. L. Merlino and N. D'Angelo, Phys. Plasmas, 2: 3563 (1997).

[22] R. Silva, Jr. A. R. Plastino and J. A. S. Lima, Phys. Lett. A, 249: 401 (1998).

[23] S. N. Chow and J. K. Hale, Method of Bifurcation Theory, Springer-Verlag, New York (1981).

[24] J. Guckenheimer and P. J. Holmes, Nonlinear Oscillations, Dynamical Systems and Bifurcations of Vector Fields, Springer-Verlag, New York (1983) (page numbers 117–156 and 289–411).

[25] Z. Fu, S. Liu, S. Liu and Q. Zhao, Physics Letters A, 290: 72 (2001).

[26] P. K. Shukla, A. A. Mamun and D. A. Mendis, Phys. Rev. E, 84: 026405 (2011).

[27] E. Wigner, Phys. Rev., 46: 1002 (1934).

[28] A. Shahzad and M. G. He, Plasma Sci. Technol., 14: 771 (2012).

[29] H. Thomas, G. E. Morfill, V. Demme, J. Goree, B. Feuerbacher and D. Mohlmann, Phys. Rev. Lett., 73: 652 (1994).

[30] E. A. Maree and G. F. Sarafanov, Phys. Plasmas, 5: 1563 (1998).

[31] M. S. Murillo, Phys. Rev. Lett., 85: 2514 (2000).

[32] M. Rosenberg and G. Kalman, Phys. Rev. E, 56: 7166 (1997).

[33] J. B. Pieper and J. Goree, Phys. Rev. Lett., 77: 3137 (1996).

[34] P. K. Kaw, Physics of Plasmas, 9: 387 (2002).

[35] J. Pramanik, G. Prasad, A. Sen and P. K. Kaw, Phys. Rev. Lett., 88: 175001 (2002).

[36] P. Bandyopadhyay, G. Prasad, A. Sen and P. K. Kaw, Physics Letters A, 368: 491 (2007).

[37] P. K. Shukla and I. Lin, Physics Letters A, 374: 1165 (2010).

[38] H. Alinejad, Physics Letters A, 374: 1855 (2010).

[39] A. P. Misra and S. Banerjee, Physical Review E, 83: (2011).

[40] S. A. Ema, M. R. Hossen and A. A. Mamun, Contrib. Plasma Phys., 55: 596 (2015).

[41] M. A. El-Borie and A. Atteya, Physics of Plasmas, 24: 113706 (2017),

[42] S. Almutalk, S. A. El-Tantawy, E. I. El-Awady and S. K. El-Labany, Physics Letters A, 383: 1937 (2019).

[43] P. K. Shukla and A. A. Mamun, IEEE Transactions on Plasma Science, 29: 22 (2001),

[44] S. Ichimaru and S. Tanaka, Phys. Rev. Lett., 56: 2815 (1986).

[45] D. Pines and P. Nozieres, The Theory of Quantum Liquids, Taylor and Francis Group, Vol. I (1966).

[46] W. L. Slattery, G. D. Doolen and H. E. DeWitt, Phys. Rev. A. Gen. Phys., 21: 2087 (1980).

[47] M. A. Berkovsky, Phys. Lett. A, 166: 365 (1992).

[48] H. Washimi and T. Taniuti, Phys. Rev. Lett., 17: 996 (1966).

[49] C. B. Dwivedi and B. P. Pandey, Physics of Plasmas, 2: 4134–4139 (1995).

[50] A. A. Mamun, Phys. Lett. A, 372: 4610 (2008).

[51] A. M. Wazwaz, Applied Mathematics and Computation, 154: 713 (2004).

[52] A. R. Seadawy, Pramana-J. Phys., 89: 49 (2017).

[53] H. Derfler and T. C. Simonen, Phys. Fluids, 12: 269 (1969).

[54] S. Ikezawa and Y. Nakamura, J. Phys. Soc. Jpn., 50: 962 (1981).

[55] B. D. Fried and R. W. Gould, Phys. Fluids, 4: 139 (1961).

[56] M. Yu and P. K. Shukla, J. Plasma Phys., 29: 409 (1983).

[57] S. P. Gary and R. L. Tokar, Phys. Fluids, 28: 2439 (1985).

[58] M. P. Dell, I. M. A. Geldhill and M. A. Hellberg, Zeitschrift für Naturforschung A, 42: 1175 (1987).

[59] N. Dubouloz, R. Pottelette, M. Malingre and R. M. Treumann, Geophys. Res. Lett., 18: 155 (1991).

[60] R. L. Mace, S. Baboolal, R. Bharuthram and M. A. Hellberg, J. Plasma Phys., 45: 323 (1991).

[61] M. Berthomier, R. Pottelette, M. Malingre and Y. Khotyainsev, Phys. Plasmas, 7: 2987 (2000).

[62] S. S. Ghosh, A. Sen and G. S. Lakhina, Nonlinear Process. Geophys., 9: 463 (2002).

[63] W. Masood and H. A. Shah, J. Fusion Energy, 22: 201 (2003).

[64] R. L. Mace and M. A. Hellberg, Phys. Plasmas, 8: 2649 (2001).

[65] N. Dubouloz, R. A. Treumann, R. Pottelette and M. Malingre, J. Geophys. Res. Atmos., 98: 17415 (1993).

[66] A. A. Mamun, P. K. Shukla and L. Stenflo, Phys. Plasmas, 9: 1474 (2002).

[67] B. Bezzerides, D. W. Forslund and E. L. Lindman, Phys. Fluids, 21: 2179 (1978).

[68] R. L. Tokar and S. P. Gary, Geophys. Res. Lett., 11: 1180 (1984).

[69] D. Schriver and M. Ashour-Abdalla, Geophys. Res. Lett., 16: 899 (1989).

[70] S. Mahmood and W. Masood, Phys. Plasmas, 15: 122302 (2008).

Chapter 5

Bifurcation of Arbitrary Amplitude Waves in Plasmas

5.1 Introduction

Bifurcation of arbitrary amplitude waves is studied in different plasma systems. A nonlinear wave with amplitude less than unity is called a small amplitude nonlinear wave. If there is no restriction on the amplitude of the nonlinear waves, then the waves are known as arbitrary amplitude waves. There are different kinds of arbitrary amplitude nonlinear waves in plasmas, for example,

- Arbitrary amplitude ion-acoustic wave.

- Arbitrary amplitude dust-ion-acoustic wave.

- Arbitrary amplitude dust-acoustic wave.

- Arbitrary amplitude electron-acoustic wave, etc.

5.2 Bifurcation of ion-acoustic waves with arbitrary amplitude

Firstly, bifurcation behavior of ion-acoustic (IA) waves with arbitrary amplitude is presented in a two-component plasma consisting of cold mobile ions and Maxwell distributed electrons.

5.2.1 Basic equations

A two-component plasma [1]-[2] consisting of cold mobile ions and Maxwell distributed electrons is considered. The non-dimensional form of basic equations is as follows:

$$\frac{\partial n}{\partial t} + \frac{\partial (nu)}{\partial x} = 0, \tag{5.1}$$

$$\frac{\partial u}{\partial t} + u\frac{\partial u}{\partial x} = -\frac{\partial \phi}{\partial x}, \tag{5.2}$$

$$\frac{\partial^2 \phi}{\partial x^2} = n_e - n, \tag{5.3}$$

where

$$n_e(\phi) = e^{\phi}. \tag{5.4}$$

Here, n denotes number density of ions and is normalized by n_0, and n_e signifies number density of electrons and is normalized by n_{e0}. In the basic equations above, u and ϕ are ion velocity and electrostatic potential, which are normalized by IA speed $c_s = (K_B T_e/m)^{1/2}$ and $K_B T_e/e$, respectively. Here, e is the charge of electron, m is mass of ions and K_B is Boltzmann's constant. The time t and space variable x are normalized respectively by $\omega^{-1} = (m/4\pi e^2 n_0)^{1/2}$ and $\lambda = (K_B T_e/4\pi e^2 n_0)^{1/2}$, where ω represents ion plasma frequency and λ denotes Debye length.

5.2.2 Formation of dynamical system and phase portraits

A traveling wave transformation $\xi = x - vt$ is considered, where v is speed of the traveling wave. Employing the transformation ξ, one can transform basic equations into a planar dynamical system. Using variable ξ and the initial condition $u = 0$, $n = 1$ and $\phi = 0$ in equations (5.1) and (5.2), one can easily obtain

$$n = \frac{v}{\sqrt{v^2 - 2\phi}}. \tag{5.5}$$

Substituting equations (5.4) and (5.5) in equation (5.3) and considering the terms of ϕ up to second degree, one can obtain

$$\frac{d^2\phi}{d\xi^2} = a\phi + b\phi^2, \tag{5.6}$$

where $a = 1 - \frac{1}{v^2}$, and $b = \frac{1}{2} - \frac{3}{2v^4}$.

Then, equation (5.6) can be represented in the form of dynamical system as:

$$\begin{cases} \frac{d\phi}{d\xi} = z, \\ \frac{dz}{d\xi} = a\phi + b\phi^2. \end{cases} \tag{5.7}$$

The system (5.7) is a planar Hamiltonian system with the following Hamiltonian function:

$$H(\phi, z) = \frac{z^2}{2} - a\frac{\phi^2}{2} - b\frac{\phi^3}{3} = h, \text{say.} \tag{5.8}$$

The planar dynamical system (5.7) has a parameter v. It is to note that the phase trajectories defined by the vector fields of equation (5.7) will determine all traveling wave solutions of equation (5.6). Now, one can study the bifurcation behavior of phase portraits of system (5.7) in the phase plane (ϕ, z) for different values of parameter v. Solitary and periodic wave solutions of equation (5.6) correspond to homoclinic and periodic trajectories of system (5.7).

Now, one can investigate the bifurcation behavior and phase portraits of the Hamiltonian system (5.7). Let $E_i(\phi_i, 0)$ to be an equilibrium point of system (5.7), there exist two equilibrium points which occur at $E_0(\phi_0, 0)$, and $E_1(\phi_1, 0)$, where $\phi_0 = 0$, $\phi_1 = -\frac{a}{b}$. Suppose $M(\phi_i, 0)$ is the coefficient matrix of the linearized system of equation (5.8) at an equilibrium point $E_i(\phi_i, 0)$, then

$$J = det M(\phi_i, 0) = -a - 2b\phi_i. \tag{5.9}$$

Implementing the theory of planar dynamical systems ([3]-[4]), an equilibrium point $E_i(\phi_i, 0)$ of the planar dynamical system is a saddle point when $J < 0$, center when $J > 0$ and cusp when $J = 0$ with Poincaré index 0 of the equilibrium point.

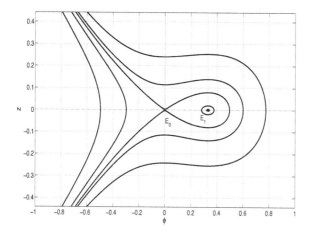

Figure 5.1: Phase portrait of system (5.7) for $v = 1.1$.

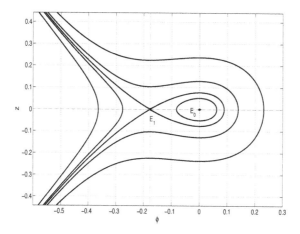

Figure 5.2: Phase portrait of system (5.7) for $v = 0.8$.

Applying the above analysis, one can obtain different phase portraits of equation (5.7) depending on parameter v, shown in Figures 5.1-5.4.

(i) When $a > 0$ and $b < 0$, the system (5.7) has two equilibrium points at $E_0(0,0)$ and $E_1(\phi_1,0)$ with $\phi_1 > 0$, where $E_1(\phi_1,0)$ is a center, and $E_0(0,0)$ is a saddle point. There is a homoclinic trajectory at $E_0(0,0)$ enclosing the center at $E_1(\phi_1,0)$ (see Figure 5.1).

(ii) When $a < 0$ and $b < 0$, the system (5.7) has two equilibrium points which occur at $E_0(0,0)$ and $E_1(\phi_1,0)$ with $\phi_1 < 0$, where $E_0(0,0)$ is a

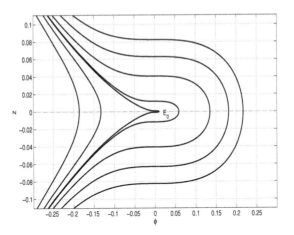

Figure 5.3: Phase portrait of system (5.7) for $v = 1$.

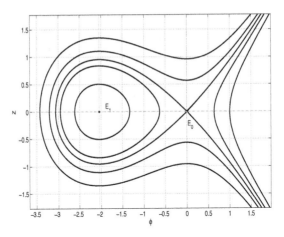

Figure 5.4: Phase portrait of system (5.7) for $v = 1.7$.

center, and $E_1(\phi_1, 0)$ is a saddle point. There is a homoclinic trajectory at $E_1(\phi_1, 0)$ enclosing the center at $E_0(0,0)$ (see Figure 5.2).

(iii) When $a = 0$ and $b < 0$, the system (5.7) has only one equilibrium point at $E_0(0,0)$, where $E_0(0,0)$ is a cusp (see Figure 5.3).

(iv) When $a > 0$ and $b > 0$, the system (5.7) has two equilibrium points at $E_0(0,0)$ and $E_1(\phi_1, 0)$ with $\phi_1 < 0$. In this case, $E_0(0,0)$ is a saddle point, and $E_1(\phi_1, 0)$ is a center. There is a homoclinic trajectory at $E_0(0,0)$ enclosing the center at $E_1(\phi_1, 0)$ (see Figure 5.4).

5.2.3 Wave solutions

In this section, the planar dynamical system (5.7) and the Hamiltonian function (5.8) are used with the condition $h = 0$ to find the solitary wave solution of equation (5.6) depending on different parametric conditions.

(i) When $a > 0$ and $b < 0$ (see Figure 5.1), the system (5.7) has compressive solitary wave solution given by

$$\phi = -\frac{3a}{2b}\operatorname{sech}^2\left(\frac{1}{2}\sqrt{a}\xi\right). \tag{5.10}$$

Effect of parameter v on the compressive solitary wave solution is shown in Figure 5.5.

(ii) When $a > 0$ and $b > 0$ (see Figure 5.4), the system (5.7) has a rarefactive solitary wave solution given by

$$\phi = -\frac{3a}{2b}\operatorname{sech}^2\left(\frac{1}{2}\sqrt{a}\xi\right). \tag{5.11}$$

Effect of parameter v on the rarefactive solitary wave solution is depicted in Figure 5.6.

The numerical wave solutions for periodic trajectories shown in Figures 5.1 and 5.2 are presented in Figures 5.7 and 5.8, respectively. These Figures 5.7-5.8 also depict the crucial effect of traveling wave v in modifying the basic properties of periodic waves.

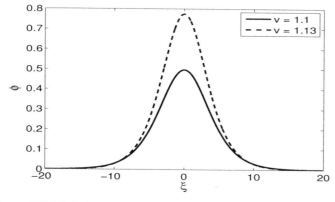

Figure 5.5: Variation of solitary waves (5.10) for different values of v.

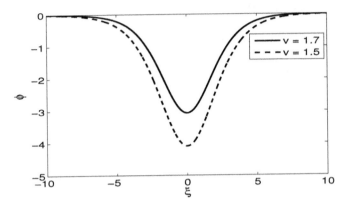

Figure 5.6: Variation of solitary waves (5.11) for different values of v.

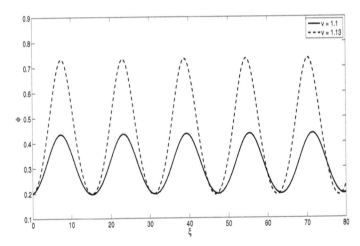

Figure 5.7: Variation of periodic waves due to v for $a > 0$ and $b < 0$.

5.3   Bifurcation of dust-ion-acoustic waves with arbitrary amplitude

5.3.1   *Basic equations*

A three-component unmagnetized plasma comprising of cold inertial ions, negatively charged stationary dust and non-inertial q-nonextensive electrons is considered. At equilibrium condition, $n_{e0} = n_0 - Z_d n_{d0}$, where Z_d denotes number of electrons residing onto the dust grain surface and n_0, n_{e0} and n_{d0} denote the equilibrium ion, electron, and dust number densities respectively.

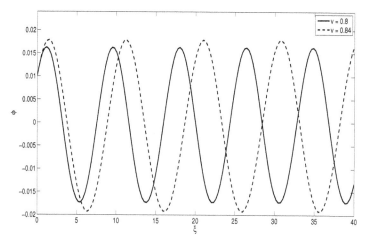

Figure 5.8: Variation of periodic waves due to v for $a > 0$ and $b > 0$.

The dynamics of dust ion acoustic waves (DIAWs), whose phase speed is much smaller than the thermal speed of electron and larger than the thermal speed of ion, is characterized by the following non-dimensional form of basic equations [5]:

$$\frac{\partial n}{\partial t} + \frac{\partial (nu)}{\partial x} = 0, \tag{5.12}$$

$$\frac{\partial u}{\partial t} + u\frac{\partial u}{\partial x} = -\frac{\partial \phi}{\partial x}, \tag{5.13}$$

$$\frac{\partial^2 \phi}{\partial x^2} = (1 - \mu)n_e - n + \mu, \tag{5.14}$$

where $\mu = \frac{Z_d n_{d0}}{n_0}$. The following nonextensive distribution function can be considered to model nonextensive electron particles [5]

$$f_e(v) = C_q\{1 + (q-1)[\frac{m_e v^2}{2k_B T_e} - \frac{e\phi}{k_B T_e}]\}^{\frac{1}{(q-1)}},$$

where ϕ is the electrostatic potential and the other variables or parameters obey their usual meaning. The function $f_e(v)$ is the special distribution which maximizes the Tsallis entropy and, therefore, conforms to the laws of thermodynamics. The constant of normalization is given by

$$C_q = n_{e0}\frac{\Gamma(\frac{1}{1-q})}{\Gamma(\frac{1}{1-q} - \frac{1}{2})}\sqrt{\frac{m_e(1-q)}{2\pi k_B T_e}} \text{ for } -1 < q < 1,$$

and

$$C_q = n_{e0} \frac{1+q}{2} \frac{\Gamma(\frac{1}{q-1}+\frac{1}{2})}{\Gamma(\frac{1}{q-1})} \sqrt{\frac{m_e(q-1)}{2\pi k_B T_e}} \text{ for } q > 1.$$

Integrating $f_e(v)$ over all velocity spaces, one can obtain the following nonextensive electron number density:

$$n_e(\phi) = n_{e0}\{1+(q-1)\frac{e\phi}{k_B T_e}\}^{1/(q-1)+1/2}.$$

Thus, the normalized electron number density [5] is obtained as:

$$n_e(\phi) = \{1+(q-1)\phi\}^{1/(q-1)+1/2}. \tag{5.15}$$

where, the parameter q is a number greater than -1, and it signifies the strength of nonextensivity.

In the equations above (5.12)-(5.15), n and n_e denote the number densities of the ions and electrons, respectively, which are normalized by equilibrium ion density n_0. Ion fluid velocity u and electrostatic potential ϕ are normalized by the ion acoustic speed $C_s = (K_B T_e/m)^{1/2}$ and $k_B T_e/e$, respectively. Here, e is the electron charge, m is the mass of ions and k_B is the Boltzmann constant. The time t and space variable x are normalized by the inverse of ion plasma frequency $\omega_p^{-1} = (m/4\pi e^2 n_0)^{1/2}$ and the Debye length $\lambda = (k_B T_e/4\pi e^2 n_0)^{1/2}$, respectively.

5.3.2 Formation of dynamical system and phase portraits

To analyze the phase portraits and find all traveling wave solutions, the traveling wave frame $\xi = x - vt$, is considered, where v is speed of the frame. Using this transformation and the initial conditions $u = 0$, $n = 1$ and $\phi = 0$, the following relation is obtained from equations (5.12) and (5.13):

$$n = \frac{v}{\sqrt{v^2 - 2\phi}}. \tag{5.16}$$

Substituting equations (5.15) and (5.16) into equation (5.14) and considering the terms of ϕ up to third degree, one can obtain

$$\frac{d^2\phi}{d\xi^2} = a\phi + b\phi^2 + c\phi^3, \tag{5.17}$$

where $\quad a = \frac{(1-\mu)(1+q)}{2} - \frac{1}{v^2}, \quad b = \frac{(1-\mu)(1+q)(3-q)}{8} - \frac{3}{2v^4}, \quad$ and
$c = \frac{(1-\mu)(1+q)(3-q)(5-3q)}{48} - \frac{5}{2v^6}.$

Then, equation (5.17) can be written equivalently in the form following dynamical system:

$$\begin{cases} \frac{d\phi}{d\xi} = z, \\ \frac{dz}{d\xi} = a\phi + b\phi^2 + c\phi^3, \end{cases} \qquad (5.18)$$

The system (5.18) is a planar Hamiltonian system with the following Hamiltonian function:

$$H(\phi,z) = \frac{z^2}{2} - a\frac{\phi^2}{2} - b\frac{\phi^3}{3} - c\frac{\phi^4}{4} = h, \text{say.} \qquad (5.19)$$

The planar dynamical system (5.17) has the parameters q, μ and v. Importantly, the phase trajectories defined by the vector fields of equation (5.18) determine all traveling wave solutions of equation (5.17). One can study the bifurcations of phase portraits of equation (5.17) in the (ϕ, z) phase plane depending on the parameters q, μ and v. A solitary wave solution of equation (5.17) corresponds to a homoclinic trajectory of system (5.18) and a periodic trajectory of equation (5.18) corresponds to a periodic traveling wave solution of equation (5.17).

Now, one can study the bifurcation set and phase portraits of the planar Hamiltonian system (5.18). It is clear that on the (ϕ, z) phase plane, the abscissas of equilibrium points of system (5.18) are the zeros of $f(\phi) = \phi(\phi^2 + \frac{b}{c}\phi + \frac{a}{c})$. Let $E_i(\phi_i, 0)$ be an equilibrium point of the dynamical system (5.18), where $f(\phi_i) = 0$. When $b^2 - 4ac > 0$, there exist three equilibrium points at $E_0(\phi_0, 0)$, $E_1(\phi_1, 0)$ and $E_2(\phi_2, 0)$, where $\phi_0 = 0$, $\phi_1 = \frac{-b+\sqrt{b^2-4ac}}{2c}$ and $\phi_2 = \frac{-b-\sqrt{b^2-4ac}}{2c}$. Considering $M(\phi_i, 0)$ as the coefficient matrix of the linearized system of the traveling system (5.18) at an equilibrium point $E_i(\phi_i, 0)$, one can get

$$J = detM(\phi_i, 0) = -cf'(\phi_i). \qquad (5.20)$$

Using the concept of planar dynamical systems ([3]-[4]), the equilibrium point $E_i(\phi_i, 0)$ of the planar dynamical system (5.18) is a saddle point when $J < 0$, and the equilibrium point $E_i(\psi_i, 0)$ of the planar dynamical system (5.18) is a center when $J > 0$.

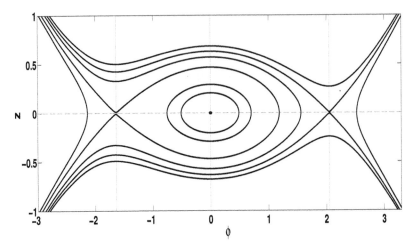

Figure 5.9: Phase portrait of system (5.18) for $q = -0.8, \mu = 0.2$ and $v = 1.98$.

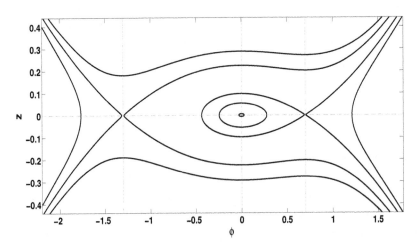

Figure 5.10: Phase portrait of system (5.18) for $q = 0.2, \mu = 0.5$ and $v = 1.7$.

Employing the systematic analysis of parameters q, μ and v, the phase portraits of the system (5.18) are presented in Figures 5.9-5.13.

(i) When $c > 0, ac < 0, bc < 0, b^2 - 4ac > 0$ and $2b^2 - 9ac > 0$, then the system (5.18) has three equilibrium points at $E_0(\phi_0, 0)$, $E_1(\phi_1, 0)$ and $E_2(\phi_2, 0)$ with $\phi_2 < 0 < \phi_1$, where $E_1(\phi_1, 0)$ and $E_2(\phi_2, 0)$ are saddle

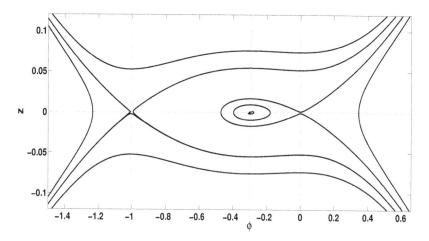

Figure 5.11: Phase portrait of system (5.18) for $q = 0.43$, with μ and v same as Figure 5.10.

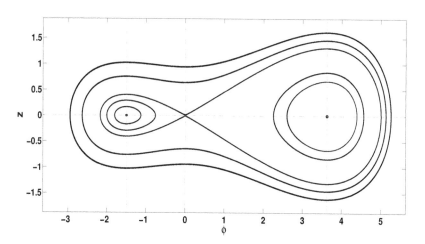

Figure 5.12: Phase portrait of system (5.18) for $q = 1.1$, with μ and v same as Figure 5.10.

points and $E_0(\phi_0, 0)$ is a center. There is a homoclinic trajectory at $E_2(\phi_2, 0)$ enclosing the center at $E_0(\phi_0, 0)$ (see Figure 5.9).

(ii) When $c > 0, ac < 0, bc > 0, b^2 - 4ac > 0$ and $2b^2 - 9ac > 0$, then the system (5.18) has three equilibrium points at $E_0(\phi_0, 0)$, $E_1(\phi_1, 0)$ and $E_2(\phi_2, 0)$ with $\phi_2 < 0 < \phi_1$, where $E_1(\phi_1, 0)$ and $E_2(\phi_2, 0)$ are saddle

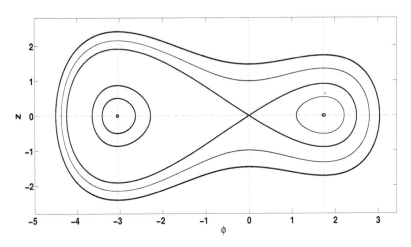

Figure 5.13: Phase portrait of system (5.18) for $q = 2.9$, with μ and ν same as Figure 5.10.

points and $E_0(\phi_0, 0)$ is a center. There is a homoclinic trajectory at $E_1(\phi_1, 0)$ enclosing the center at $E_0(\phi_0, 0)$ (see Figure 5.10).

(iii) When $c > 0, ac > 0, bc > 0, b^2 - 4ac > 0$ and $2b^2 - 9ac > 0$, then the system(5.18) has three equilibrium points at $E_0(\phi_0, 0)$, $E_1(\phi_1, 0)$ and $E_2(\phi_2, 0)$ with $\phi_2 < \phi_1 < 0$, where $E_0(\phi_0, 0)$ and $E_2(\phi_2, 0)$ are saddle points and $E_1(\phi_1, 0)$ is a center. There is a homoclinic trajectory at $E_0(\phi_0, 0)$ enclosing the center at $E_1(\phi_1, 0)$ (see Figure 5.11).

(iv) When $c < 0, ac < 0, bc < 0, b^2 - 4ac > 0$ and $2b^2 - 9ac > 0$, then the system (5.18) has three equilibrium points at $E_0(\phi_0, 0)$, $E_1(\phi_1, 0)$ and $E_2(\phi_2, 0)$ with $\phi_1 < 0 < \phi_2$, where $E_1(\phi_1, 0)$ and $E_2(\phi_2, 0)$ are centers and $E_0\phi_0, 0)$ is a saddle point. There is a pair of homoclinic trajectories at $E_0(\phi_0, 0)$ enclosing the centers at $E_1(\phi_1, 0)$ and $E_2(\phi_2, 0)$ (see Figure 5.12).

(v) When $c < 0, ac < 0, bc > 0, b^2 - 4ac > 0$ and $2b^2 - 9ac > 0$, then the system (5.18) has three equilibrium points at $E_0(\phi_0, 0)$, $E_1(\phi_1, 0)$ and $E_2(\phi_2, 0)$ with $\phi_1 < 0 < \phi_2$, where $E_1(\phi_1, 0)$ and $E_2(\phi_2, 0)$ are centers and $E_0\phi_0, 0)$ is a saddle point. There is a pair of homoclinic trajectories at $E_0(\phi_0, 0)$ enclosing the centers at $E_1(\phi_1, 0)$ and $E_2(\phi_2, 0)$ (see Figure 5.13).

5.3.3 Wave solutions

Employing the concept of dynamical system (5.18) and the Hamiltonian function (5.19), three forms of traveling wave solutions (solitary and periodic wave solutions) are derived. It is important to note that $sn(\Omega\xi,k)$ is the Jacobian elliptic function [6] with the modulo k.

(i) Corresponding to a family of periodic trajectories enclosing an equilibrium point $E_0(\phi_0,0)$ in Figure 5.9, system (5.18) has a family of periodic wave solutions which can be obtained by:

$$\phi(\xi) = \frac{(\beta_1 - \gamma_1)\delta_1 sn^2(\Omega\xi,k) - \gamma_1(\beta_1 - \delta_1)}{(\beta_1 - \gamma_1)sn^2(\Omega\xi,k) - (\beta_1 - \delta_1)}, \qquad (5.21)$$

where, $\Omega = \sqrt{-\frac{c}{8}(\beta_1 - \delta_1)(\gamma_1 - \alpha_1)}$, $k = \sqrt{\frac{(\alpha_1 - \delta_1)(\beta_1 - \gamma_1)}{(\alpha_1 - \gamma_1)(\beta_1 - \delta_1)}}$, where $\alpha_1, \beta_1, \gamma_1$ and δ_1 are solutions of the equation $h + \frac{c}{4}\phi^4 + \frac{b}{3}\phi^3 + \frac{a}{2}\phi^2 = 0$, with $\alpha_1 > \beta_1 > \gamma_1 > \delta_1$, $h \in (h_2,0)$.

(ii) Corresponding to homoclinic trajectory at $E_0(\phi_0,0)$ in Figure 5.11 at $E_0(\phi_0,0)$, system (5.18) has solitary wave solution which can be obtained by:

$$\phi(\xi) = \frac{(\frac{4b}{c} + 6A\sqrt{\frac{2a}{c}})}{3(A^2 - 1)}, \qquad (5.22)$$

where $A = exp(\mp\sqrt{a}\xi) - \frac{b}{3}\sqrt{\frac{2}{ac}}$.

(iii) Corresponding to a pair of homoclinic trajectories at $E_0(\phi_0,0)$ in Figure 5.12, system (5.18) has both compressive and rarefactive solitary wave solutions:

$$\phi(\xi) = \pm\frac{1}{\sqrt{2(1 - \frac{b^2}{9ac})}\, sin(2\sqrt{\frac{a}{c}}\xi) + \frac{b}{6a}}. \qquad (5.23)$$

Note that corresponding to homoclinic and periodic trajectories of the other phase portraits (Figures 5.10 and 5.13) of the system (5.18), one can also obtain the solitary and periodic wave solutions.

5.4 Bifurcation of dust-acoustic waves with arbitrary amplitude

Bifurcation behavior of dust-acoustic waves with arbitrary amplitude is presented in a three-component unmagnetized plasma comprising of negatively charged inertial dust particles, κ-distributed electrons and ions.

5.4.1 Basic equations

The dynamics nonlinear dust-acoustic waves with low-phase velocity is described by the following normalized form of basic equations [2]:

$$\frac{\partial n}{\partial t} + \frac{\partial (nu)}{\partial x} = 0, \tag{5.24}$$

$$\frac{\partial u}{\partial t} + u\frac{\partial u}{\partial x} = \frac{\partial \phi}{\partial x}, \tag{5.25}$$

$$\frac{\partial^2 \phi}{\partial x^2} = n + n_e - n_i. \tag{5.26}$$

The densities of electrons and ions are given by the following of κ-distribution function:

$$n_e = \frac{1}{p-1}\left(1 - \sigma\frac{\phi}{\kappa - 3/2}\right)^{-\kappa+1/2}, \tag{5.27}$$

$$n_i = \frac{p}{p-1}\left(1 + \frac{\phi}{\kappa - 3/2}\right)^{-\kappa+1/2}. \tag{5.28}$$

In the above equations (5.24)-(5.28), n, n_e, and n_i denote the number densities of the inertial dust particles, electrons and ions, respectively, which are normalized by their unperturbed densities. In this case, the dust fluid velocity u and electrostatic potential ϕ are normalized by the dust acoustic speed $c = (ZT_i/m)^{1/2}$, and T_e/e, respectively, where m is the mass of dust particles and Z is the number of charges residing on the dust grain surface. The time and space variables are normalized to the dust plasma period $\omega^{-1} = (m/4\pi n_0 Ze^2)^{1/2}$ and the Debye length $\lambda = (T_i/4\pi n_0 e^2)^{1/2}$, respectively. Here, $p = \frac{n_{i0}}{n_{e0}}$, and $\sigma = \frac{T_i}{T_e}$.

5.4.2 Formation of dynamical system and phase portraits

For a localized solution, one can introduce a frame $\xi = x - vt$, where v is the velocity of the frame. Adopting ξ and using the initial condition $u = 0, n = 1$

and $\phi = 0$ in equations (5.24) and (5.25), the following relation for number density of dust particles can be obtained

$$n = \frac{v}{\sqrt{v^2 + 2\phi}}.$$
(5.29)

Using equations (5.27), (5.28) and (5.29) into equation (5.26) and neglecting the higher order (more than second degree) terms of ϕ, one can have

$$\frac{d^2\phi}{d\xi^2} = a\phi^2 + b\phi.$$
(5.30)

where $a = \frac{(\kappa^2 - 1/4)(\sigma^2 - p)}{2(p-1)(\kappa - 3/2)^2} + \frac{3}{2v^4}$ and $b = \frac{(\kappa - 1/2)(\sigma + p)}{(p-1)(\kappa - 3/2)} - \frac{1}{v^2}$.

Then, equation (5.30) is equivalent to the following system:

$$\begin{cases} \frac{d\phi}{d\xi} = z, \\ \frac{dz}{d\xi} = a\phi^2 + b\phi. \end{cases}$$
(5.31)

The system (5.31) is a planar Hamiltonian system with the Hamiltonian function:

$$H(\phi, z) = \frac{z^2}{2} - a\frac{\phi^3}{3} - b\frac{\phi^2}{2}.$$
(5.32)

The system (5.32) is a dynamical system with plasma parameters $\kappa, \sigma, p,$ and v.

All probable periodic and homoclinic trajectories defined by the vector field equation (5.31) are obtained when the plasma parameters κ, σ, p and v are varied. When $ab \neq 0$, there exist two equilibrium points at $E_0(\phi_0, 0)$ and $E_1(\phi_1, 0)$, with $\phi_0 = 0$ and $\phi_1 = -\frac{b}{a}$. If $M(\phi_i, 0)$ denotes the coefficient matrix of the linearized system of equation (5.31) at an equilibrium point $E_i(\phi_i, 0)$, then the determinant of the Jacobian matrix is given by

$$J = detM(\phi_i, 0) = -b - 2a\phi_i.$$
(5.33)

Using the theory of planar dynamical systems ([3]-[4]), an equilibrium point $E_i(\phi_i, 0)$ of the dynamical system is a saddle point when $J < 0$ and center when $J > 0$.

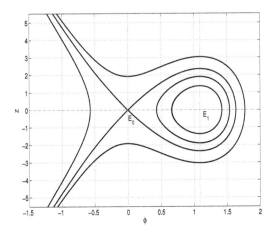

Figure 5.14: Phase portrait of system (5.31) for $\kappa = 1.8, \sigma = 0.1, p = 1.3$ and $v = 0.4$.

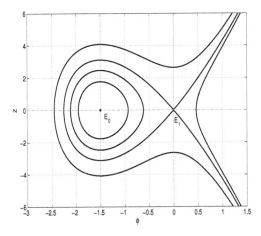

Figure 5.15: Phase portrait of system (5.31) for $\kappa = 1.8, \sigma = 0.1, p = 1.3$ and $v = 0.37$.

Applying the analysis above, one can obtain all distinct phase portraits of equation (5.31) on varying the system's parameters, shown in Figures 5.14-5.15.

For the phase portrait given by Figure 5.14, the parameters κ, σ, p, and v satisfy the relations $a < \frac{3}{2v^4}$ and $b < \frac{1}{v^2}$. For this case, one can get homoclinic trajectory at the equilibrium point $E_0(\phi_0, 0)$ and a family of periodic trajectories around $E_1(\phi_1,0)$. Here, $E_0(\phi_0,0)$ is a saddle point and $E_1(\phi_1,0)$ is a center. For the phase portrait given by Figure 5.15, the parameters κ, σ, p, and v satisfy the relations $a < \frac{3}{2v^4}$ and $b > \frac{1}{v^2}$. Now, one can get homoclinic trajectory at the equilibrium point $E_1(\phi_1,0)$ and a family of periodic trajectories around $E_0(\phi_0,0)$. The equilibrium point $E_0(\phi_0,0)$ is a center and $E_1(\phi_1,0)$ is a saddle point.

5.4.3 Wave solutions

With the help of planar dynamical system (5.31) and the Hamiltonian function equation (5.32) with $h = 0$, one can find the solitary wave solution of equation (5.30), depending on different parametric conditions.

(i) When $a < 0$ and $b > 0$ (see Figure 5.4.2), the equation (5.30) has a compressive solitary wave solution given by

$$\phi = -\frac{3b}{2a}sech^2(\frac{1}{2}\sqrt{b}\xi). \qquad (5.34)$$

Variations of compressive solitary waves for different values of κ and p are displayed in Figures 5.16 and 5.17, respectively.

(ii) When $a > 0$ and $b > 0$ (see Figure 5.15), equation (5.30) has a rarefactive solitary wave solution which is given by

$$\phi = -\frac{3b}{2a}sech^2(\frac{1}{2}\sqrt{b}\xi). \qquad (5.35)$$

Variations of rarefactive solitary waves for different values of κ and p are presented in Figures 5.18 and 5.19, respectively.

5.5 Bifurcation of electron-acoustic waves with arbitrary amplitude

A noticeable number of works have reported the propagation of electron-acoustic waves (EAWs) in plasma literature. Fundamentally, EAWs are high frequency electrostatic modes in comparison with the ion plasma frequency [7] which can propagate in unmagnetized plasma [7]-[9] as well as in magnetized plasma [10]-[19]. It is important to note that in an electrostatic wave, the inertia is provided by the cold electrons and the restoring force is given by the pressure of the hot electrons. Furthermore, ions play the role of a neutralizing background and there is no influence of the ion dynamics on the EAWs.

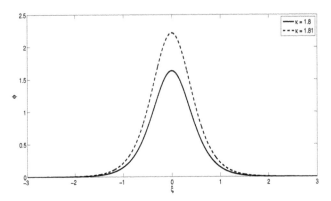

Figure 5.16: Variation of solitary waves for different values of κ with $\sigma = 0.1, p = 1.3$ and $v = 0.4$.

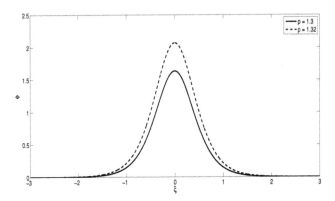

Figure 5.17: Variation of solitary waves for different values of p with $\kappa = 1.8, \sigma = 0.1$ and $v = 0.4$.

The phase speed of the EAWs lies between the cold and hot-electron thermal velocities. EA waves are influential for their excitation in space as well as in laboratory plasma, and also vital for their potential importance [13]. The idea of EAWs had been originated by Fried and Gould [14] during their numerical simulations of the linear electrostatic Vlasov dispersion equation in an unmagnetized homogeneous plasma. These waves can appear in the cusp region of the terrestrial magnetosphere [15]-[16], the geomagnetic tail [17], and the dayside auroral accelaration region [18]. Dubouloz et al. [19]-[20] studied the electron-acoustic solitons in an unmagnetized and a magnetized

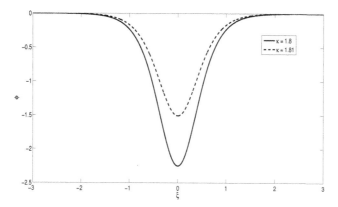

Figure 5.18: Variation of solitary waves for different values of κ with $\sigma = 0.1; p = 1.3$ and $v = 0.37$.

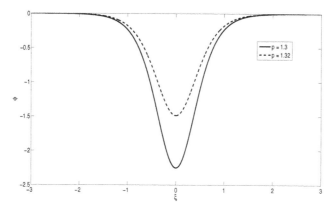

Figure 5.19: Variation of solitary waves for different values of p with $\kappa = 1.8, \sigma = 0.1$ and $v = 0.37$.

plasma and reported the fact that the negative-polarity electro-static solitary potential structures are found by the Viking satellite in the dayside auroral zone. Positive-polarity soliton structures have been studied in the auroral plasma by the FAST and POLAR spacecraft [21]. Singh et al. [16] investigated electron-acoustic solitons in four-component plasmas successfully to apply their results to Viking satellite observations in the dayside auroral zone. Lakhina et al. [22]-[23] studied electron-acoustic nonlinear solitary waves in three and four-component plasmas and applied their results to the magnetosheath plasma and the plasma sheet boundary layer.

5.5.1 Basic equations

In this work, a homogeneous, unmagnetized plasma consisting of a cold electron fluid, q-nonextensive hot electrons and stationary ions is considered. Nonlinear dynamics of the electron-acoustic waves (EAWs) in this plasma system is described by the following normalized equations [24]:

$$\frac{\partial n}{\partial t} + \frac{\partial (nu)}{\partial x} = 0, \tag{5.36}$$

$$\frac{\partial u}{\partial t} + u\frac{\partial u}{\partial x} = \alpha\frac{\partial \phi}{\partial x}, \tag{5.37}$$

$$\frac{\partial^2 \phi}{\partial x^2} = n_e + \frac{1}{\alpha}n - (1+\frac{1}{\alpha}). \tag{5.38}$$

In equations (5.36)-(5.38), n denotes the cold electron number density normalized by n_0, n_e denotes the hot electron number density, u denotes the cold electron velocity normalized by ion fluid speed $C_s = \sqrt{k_B T_h/\alpha m_e}$, where $\alpha = n_{e0}/n_0 > 1$, m_e is the electron mass, T_h is the temperature of hot electron, e is the electron charge, k_B is the Boltzmann constant, ϕ is the electrostatic potential normalized by $k_B T_h/e$. The space and time variables are normalized by the hot electron Debye radius $\lambda_h = \sqrt{k_B T_h/4\pi n_{e0}e^2}$ and the inverse of cold electron plasma frequency $\omega_{pc}^{-1} = \sqrt{m_e/4\pi n_0 e^2}$, respectively.

The normalized number density of q-nonextensive [41] hot electrons is given by

$$n_e = \{1+(q-1)\phi\}^{\frac{1}{q-1}+\frac{1}{2}}, \tag{5.39}$$

where, the parameter q indicates the strength of nonextensivity. It is important to note that when $q < -1$, the q-distribution is not normalizable [41]. The strength of nonextensivity, q varies as $-1 < q < 1$. When q = 1 the distribution function exhibits Maxwell-Boltzmann velocity distribution.

5.5.2 Formation of dynamical system and phase portraits

In this section, one can transform the model equations (5.36)-(5.38) into a planar dynamical system and present bifurcations of phase portraits of the system. Thus, taking a new variable $\xi = x - vt$, where v is speed of the electron acoustic traveling wave and substituting it into equations (5.36) and (5.37) and using the initial condition $u = 0, n = 1$ and $\phi = 0$, one can express the cold electron number density as

$$n = \frac{v}{\sqrt{v^2 + 2\alpha\phi}}. \tag{5.40}$$

Now, substituting equations (5.39), and (5.40) into equation (5.38) and considering the terms involving ϕ up to third degree, one can have

$$\frac{d^2\phi}{d\xi^2} = a\phi + b\phi^2 + c\phi^3, \tag{5.41}$$

where $a = \frac{(q+1)}{2} - \frac{1}{v^2}$, $b = \frac{(q+1)(3-q)}{8} + \frac{3\alpha}{2v^4}$, and $c = \frac{(q+1)(3-q)(5-3q)}{48} - \frac{5\alpha^2}{2v^6}$.

Then, equation (5.41) is equivalent to the following planar dynamical system:

$$\begin{cases} \frac{d\phi}{d\xi} = z, \\ \frac{dz}{d\xi} = c\phi(\phi^2 + \frac{b}{c}\phi + \frac{a}{c}). \end{cases} \tag{5.42}$$

The system (5.42) is a planar Hamiltonian system with Hamiltonian function:

$$H(\phi, z) = \frac{z^2}{2} - a\frac{\phi^2}{2} - b\frac{\phi^3}{3} - c\frac{\phi^4}{4} = h, \text{say}. \tag{5.43}$$

The system (5.42) is a dynamical system with parameters q, α and v. The phase trajectories defined by the vector fields of equation (5.42) will determine all traveling wave solutions of equation (5.41). A homoclinic and periodic trajectories of system (5.42) give solitary and periodic wave solutions of the equation (5.41), respectively.

The bifurcation theory of planar dynamical systems plays an important role ([3]-[4]) to analyze the dynamics of EAWs.

Now, one can study the bifurcation set and phase portraits of the planar Hamiltonian system (5.42). Clearly, on the (ϕ, z) phase plane, the abscissas of equilibrium points of system (5.42) are the zeros of $f(\phi) = \phi(\phi^2 + \frac{b}{c}\phi + \frac{a}{c})$. Let $E_i(\phi_i, 0)$ be an equilibrium point of the dynamical system (5.42) where $f(\phi_i) = 0$. When $b^2 - 4ac > 0$, there exist three equilibrium points at $E_0(\phi_0, 0)$, $E_1(\phi_1, 0)$ and $E_2(\phi_2, 0)$, where $\phi_0 = 0$, $\phi_1 = \frac{-b + \sqrt{b^2 - 4ac}}{2c}$ and $\phi_2 = \frac{-b - \sqrt{b^2 - 4ac}}{2c}$. If $M(\phi_i, 0)$ is the coefficient matrix of the linearized system of the traveling system (5.42) at an equilibrium point $E_i(\phi_i, 0)$, then one can get

$$J = det M(\phi_i, 0) = -cf'(\phi_i). \tag{5.44}$$

By the theory of planar dynamical systems [4], it is clear that the equilibrium point $E_i(\phi_i, 0)$ of the planar dynamical system (5.42) is a saddle point when $J < 0$ and the equilibrium point $E_i(\psi_i, 0)$ of the planar dynamical system (5.42) is a center when $J > 0$.

Applying the systematic analysis of the physical parameters q, α, and v, one can consider different cases and present all probable phase portraits of the system (5.42) in Figures 5.20-5.24.

(i) When $c < 0, ac > 0, bc < 0, b^2 - 4ac > 0$ and $2b^2 - 9ac > 0$, then the system (5.42) has three equilibrium points at $E_0(\phi_0, 0)$, $E_1(\phi_1, 0)$ and $E_2(\phi_2, 0)$ with $0 < \phi_1 < \phi_2$, where $E_0(\phi_0, 0)$ and $E_2(\phi_2, 0)$ are centers, and $E_1(\phi_1, 0)$ is a saddle point. There is a pair of homoclinic trajectories at $E_1(\phi_1, 0)$ enclosing the centers at $E_0(\phi_0, 0)$ and $E_2(\phi_2, 0)$ (see Figure 5.20).

(ii) When $c < 0, ac > 0, bc < 0$, and $b^2 - 4ac > 0$, then the system (5.42) has three equilibrium points at $E_0(\phi_0, 0)$, $E_1(\phi_1, 0)$ and $E_2(\phi_2, 0)$ with $0 < \phi_1 < \phi_2$, where $E_0(\phi_0, 0)$ and $E_2(\phi_2, 0)$ are centers, and $E_1(\phi_1, 0)$ is a saddle point. There is a pair of homoclinic trajectories at $E_1(\phi_1, 0)$ enclosing the centers at $E_0(\phi_0, 0)$ and $E_2(\phi_2, 0)$ (see Figure 5.21).

(iii) When $c < 0, ac < 0, bc < 0$, and $b^2 - 4ac > 0$, then the system (5.42) has three equilibrium points at $E_0(\phi_0, 0)$, $E_1(\phi_1, 0)$ and $E_2(\phi_2, 0)$ with $\phi_1 < 0 < \phi_2$, where $E_1(\phi_1, 0)$ and $E_2(\phi_2, 0)$ are centers, and $E_0(\phi_0, 0)$ is a saddle point. There is a pair of homoclinic trajectories at $E_0(\phi_0, 0)$ enclosing the centers at $E_1(\phi_1, 0)$ and $E_2(\phi_2, 0)$ (see Figure 5.22).

(iv) When $c > 0, ac > 0, bc > 0, b^2 - 4ac > 0$ and $2b^2 - 9ac < 0$, then the system (5.42) has three equilibrium points at $E_0(\phi_0, 0)$, $E_1(\phi_1, 0)$ and $E_2(\phi_2, 0)$ with $\phi_2 < \phi_1 < 0$, where $E_0(\phi_0, 0)$ and $E_2(\phi_2, 0)$ are saddle points, and $E_1(\phi_1, 0)$ is a center. There is a homoclinic trajectory at $E_2(\phi_2, 0)$ enclosing the center at $E_1(\phi_1, 0)$ (see Figure 5.23).

(v) When $c > 0, ac > 0, bc > 0, b^2 - 4ac > 0$ and $2b^2 - 9ac > 0$, then the system (5.42) has three equilibrium points at $E_0(\phi_0, 0)$, $E_1(\phi_1, 0)$ and $E_2(\phi_2, 0)$ with $\phi_2 < \phi_1 < 0$, where $E_0(\phi_0, 0)$ and $E_2(\phi_2, 0)$ are saddle points, and $E_1(\phi_1, 0)$ is a center. There is a homoclinic trajectory at $E_0(\phi_0, 0)$ enclosing the center at $E_1(\phi_1, 0)$ (see Figure 5.24).

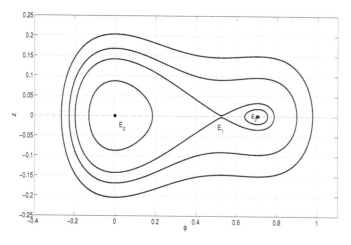

Figure 5.20: Phase portrait of system (5.42) for $q = -0.31$, $\alpha = 1.21$ and $v = 1.2$.

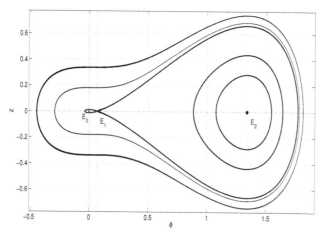

Figure 5.21: Phase portrait of system (5.42) for $q = 0.21$, $\alpha = 1.21$ and $v = 1.2$.

5.5.3 Wave solutions

In this section, the explicit electron acoustic solitary wave solutions and periodic wave solutions are obtained with the help of the dynamical system (5.42) and the Hamiltonian function (5.43). The function $sn(\Omega\xi, k)$ is the Jacobian elliptic function [6] with the modulo k.

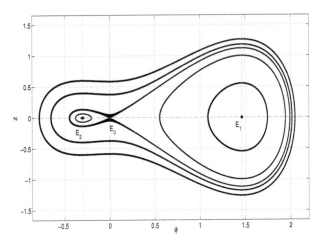

Figure 5.22: Phase portrait of system (5.42) for $q = 1.41, \alpha = 1.21$ and $v = 1.2$.

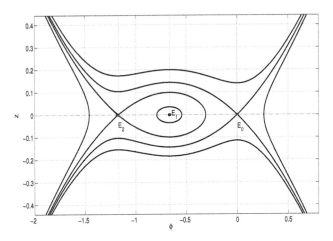

Figure 5.23: Phase portrait of system (5.42) for $q = -0.11, \alpha = 1.21$ and $v = 2$.

(i) Corresponding to a family of periodic trajectories about $E_0(\phi_0, 0)$ in Figure 5.20, system has a family of electron acoustic periodic wave solutions:

$$\phi(\xi) = \frac{(\gamma_1 - \delta_1)\beta_1 sn^2(\Omega\xi, k) - \gamma_1(\beta_1 - \delta_1)}{(\gamma_1 - \delta_1)sn(\Omega\xi, k) - (\beta_1 - \delta_1)}, \qquad (5.45)$$

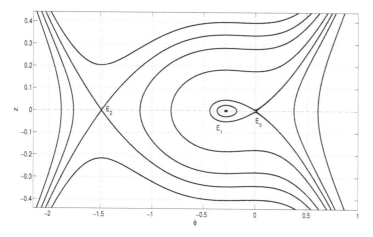

Figure 5.24: Phase portrait of system (5.42) for $q = -0.31, \alpha = 1.21$ and $v = 2$.

with $\Omega = \sqrt{-\frac{c}{8}(\beta_1 - \delta_1)(\alpha_1 - \gamma_1)}$, $k = \sqrt{\frac{(\alpha_1 - \beta_1)(\gamma_1 - \delta_1)}{(\alpha_1 - \gamma_1)(\beta_1 - \delta_1)}}$, where $\alpha_1, \beta_1, \gamma_1$ and δ_1 are roots of the equation $h + \frac{c}{4}\phi^4 + \frac{b}{3}\phi^3 + \frac{a}{2}\phi^2 = 0$, $h \in (0, h_2)$.

(ii) Corresponding to pair of homoclinic trajectories at $E_0(\phi_0, 0)$ in Figure 5.22, system (5.42) has both compressive and rarefactive electron acoustic solitary wave solutions:

$$\phi(\xi) = \pm\frac{1}{\sqrt{2(1 - \frac{b^2}{9ac})}\, sin(2\sqrt{\frac{a}{c}}\xi) + \frac{b}{6a}}. \tag{5.46}$$

(iii) Corresponding to homoclinic trajectory at $E_0(\phi_0, 0)$ in Figure 5.24, system (5.42) has electron acoustic solitary wave solution:

$$\phi(\xi) = \frac{(\frac{4b}{c} + 6A\sqrt{\frac{2a}{c}})}{3(A^2 - 1)}, \tag{5.47}$$

where $A = exp(\sqrt{a}\xi) - \frac{b}{3}\sqrt{\frac{2}{ac}}$.

References

[1] A. S. Bains, M. Tribeche and T. S. Gill, Phys. Plasmas, 18: 022108 (2011).

[2] A. Saha and P. Chatterjee, Astrophysics and Space Science, 350: 631 (2014).

[3] A. Saha, Commun. Nonlinear Sci. Numer. Simulat., 17: 3539 (2012).

[4] J. Guckenheimer and P. J. Holmes, Nonlinear Oscillations, Dynamical Systems and Bifurcations of Vector Fields, New York, Springer-Verlag (1983).

[5] A. Saha and P. Chatterjee, Eur. Phys. J. D., 69: 203 (2015).

[6] P. F. Byrd and M. D. Friedman, Handbook of Elliptic Integrals for Engineer and Scientists, Springer, New York (1971).

[7] T. H. Stix, Waves in Plasmas, AIP Publishing, New York (1992).

[8] A. A. Mamun and P. K. Shukla, J. Geophys. Res., 107: 1135 (2002).

[9] W. F. El-Taibany and W. M. Moslem, Phys. Plasmas, 12: 032307 (2005).

[10] P. K. Shukla, A. A. Mamun and B. Eliasson, Geophys. Res. Lett., 31: L07803 (2004).

[11] M. Shalaby, S. K. El-Labany, R. Sabry and L. S. El- Sherif, Phys. Plasmas, 18: 062305 (2011).

[12] M. G. Anowar and A. A. Mamun, Phys. Plasmas, 15: 102111 (2008).

[13] B. Sahu, Phys. Scr., 82: 065504 (2010).

[14] B. D. Fried and R. W. Gould, Physics of Fluids, 4: 139 (1961).

[15] R. L. Tokar and S. P. Gray, Geophys. Res. Lett., 11: 1180 (1984).

[16] S. V. Singh, R. V. Reddy and G. S. Lakhina, Adv. Space. Res., 28: 1643 (2001).

[17] D. Schriver and M. Ashour-Abdalla, Geophys. Res. Lett., 16: 899 (1989).

[18] R. Pottelette, R. E. Ergun, R. A. Treumann, M. Berthomier, C. W. Carlson, J. P. McFadden and I. Roth, Geophys. Res. Lett., 26: 2629 (1999).

[19] N. Dubouloz, R. Pottelette, M. Malingre, G. Holmgren and P. A. Lindqvist, J. Geophys. Res., 96: 3565 (1991).

[20] N. Dubouloz, R. A. Trueman, R. Pottelette and M. Malingre, J. Geophys. Res., 98: 17415 (1993).

[21] M. Berthomier, R. Pottelette, L. Muschietti, I. Roth and C. W. Carlson, Geophys. Res. Lett., 30: 2148 (2003).

[22] G. S. Lakhina, A. P. Kakad, S. V. Singh and F. Verheest, Phys. Plasmas, 15: 062903 (2008).

[23] G. S. Lakhina, S. V. Singh, A. P. Kakad, F. Verheest and R. Bharuthram, Nonlinear Processes Geophys., 15: 903 (2008a).

[24] A. Saha and P. Chatterjee, Astrophys. Space Sci., 353: 163–168 (2014).

Chapter 6

Bifurcation Analysis of Supernonlinear Waves

6.1 Introduction: supernonlinear waves

A new type of nonlinear wave in multi-component plasmas, characterized by the nontrivial topology of their phase portraits, is called supernonlinear wave (SNW) [1]. The orbits or trajectories in the phase portrait corresponding to such waves are evolved by at least two stable equilibrium points (centers) and one separatrix layer. The SNWs are possible in a plasma system, which contains three or more plasma components. The SNWs can be studied by using both reductive perturbation technique (RPT) and direct method or Sagdeev potential approach. It is to be noted that each trajectory in the phase portrait of a planar dynamical system gives a wave solution of the corresponding plasma system.

6.1.1 Different kind of trajectories

Depending upon the number of stable equilibrium points (centers) and separatrix layers in a phase plot of a dynamical system, different kinds of nonlinear and supernonlinear trajectories can be there. The following symbols are used to denote different kind of trajectories throughout this chapter:

- $\text{SNPT}_{m,n} \longrightarrow$ supernonlinear periodic trajectory,

- $\text{SNHT}_{m,n} \longrightarrow$ supernonlinear homoclinic trajectory,

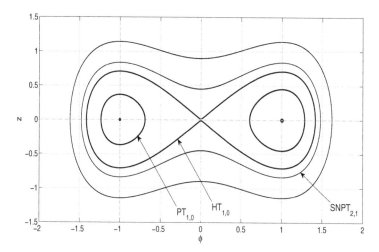

Figure 6.1: Phase portrait containing three equilibrium points with one saddle and two centers.

- $\text{PT}_{m,n} \longrightarrow$ periodic trajectory,

- $\text{HT}_{m,n} \longrightarrow$ homoclinic trajectory,

- $\text{HCT}_{m,n} \longrightarrow$ heteroclinic trajectory,

where m is the number of stable equilibrium points (centers) enveloped by the trajectory and n is the number of separatrix layers enveloped by the trajectory. A variety of nonlinear and supernonlinear trajectories is presented in Figures 6.1-6.5.

6.2 Bifurcation of supernonlinear ion-acoustic waves

6.2.1 Basic equations

A three-component plasma system composed of cold fluid ions and kappa distributed different temperature (hot and cold) electrons is considered. Propagation of IAWs is governed by the following fundamental equations [2]:

$$\frac{\partial n}{\partial t} + \frac{\partial}{\partial x}(nu) = 0, \qquad (6.1)$$

$$\frac{\partial u}{\partial t} + u\frac{\partial u}{\partial x} = -\frac{\partial \phi}{\partial x}, \qquad (6.2)$$

$$\frac{\partial^2 \phi}{\partial x^2} = f\left(1 - \frac{\alpha_c \phi}{\kappa - \frac{3}{2}}\right)^{-\kappa + \frac{1}{2}} + (1-f)\left(1 - \frac{\alpha_h \phi}{\kappa - \frac{3}{2}}\right)^{-\kappa + \frac{1}{2}} - n, \qquad (6.3)$$

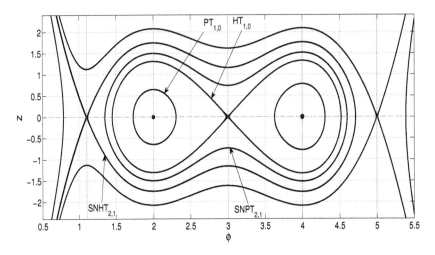

Figure 6.2: Phase portrait containing five equilibrium points with three saddles and two centers.

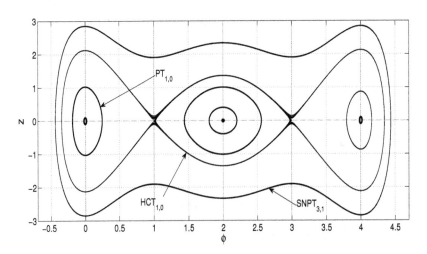

Figure 6.3: Phase portrait containing five equilibrium points with two saddles and three centers.

where n, u, ϕ and f are, respectively, number density of cold ions, velocity of ions, electrostatic potential and fractional charge density of cold electrons. $\alpha_c = \dfrac{T_{eff}}{T_c}$ and $\alpha_h = \dfrac{T_{eff}}{T_h}$, where $T_{eff} = \dfrac{T_c T_h}{f T_h + (1 - f) T_c}$ is effective temperature with hot and cold electron temperatures T_h and T_c, respectively.

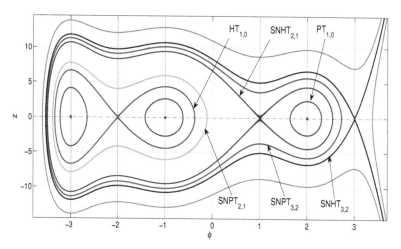

Figure 6.4: Phase portrait containing six equilibrium points with three saddles and three centers.

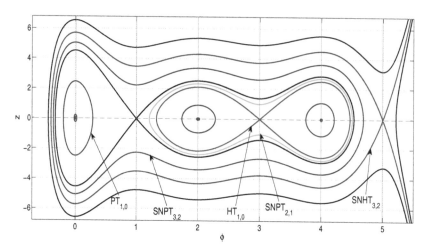

Figure 6.5: Phase portrait containing six equilibrium points with three saddles and three centers.

The considered plasma system is normalized by: n_0 normalizes n, $C_s = \left(\dfrac{k_B T_e}{m} \right)^{\frac{1}{2}}$ normalizes u, where k_B denotes the Boltzmann constant, m stands for ion mass, and e denotes strength of electron charge. Here, $\dfrac{k_B T_e}{e}$

normalizes ϕ, $\omega^{-1} = \left(\dfrac{m}{4\pi n_0 e^2}\right)^{\frac{1}{2}}$ normalizes t, where ω stands for frequency of plasma and the Debye length $\lambda_D = \left(\dfrac{k_B T_e}{4\pi n_0 e^2}\right)^{\frac{1}{2}}$ normalizes x.

6.2.2 Modified KdV equation

The KdV equation is obtained (Refer Section 4.2, Chapter 4) as

$$\frac{\partial \phi_1}{\partial \tau} + A\phi_1 \frac{\partial \phi_1}{\partial \xi} + B\frac{\partial^3 \phi_1}{\partial \xi^3} = 0, \qquad (6.4)$$

where $A = \frac{v^3}{2}\left(\frac{3}{v^4} - 2b(f\alpha_c^2 + (1-f)\alpha_h^2)\right)$ and $B = \frac{v^3}{2}$.

It is observed numerically that, for certain critical values of the physical parameters, such as $\kappa = 4.7866$, $\alpha_c = 1.1$, $\alpha_h = 0.09$ and $f = 0.2522$, the nonlinear coefficient A of the KdV equation vanishes. For such set of values, the KdV equation is not valid. Therefore, it is required to derive the mKdV equation for IAWs considering following stretching

$$\xi = \varepsilon(x - vt) \quad \text{and} \quad \tau = \varepsilon^3 t. \qquad (6.5)$$

Substitute equation (6.5) and suitable expansions of the dependent variables as the KdV equation (6.4) into fundamental equations (6.1)-(6.3) and obtain the following equations comparing the coefficients of ε^2,

$$n_2 = a\phi_2(f\alpha_c + (1-f)\alpha_h) - b\phi_2^2(f\alpha_c^2 + (1-f)\alpha_h^2), \qquad (6.6)$$

$$u_2 = \frac{1}{v}\left(\frac{\phi_1^2}{2v^2} + \phi_2\right). \qquad (6.7)$$

The following equations are deduced comparing the coefficients of ε^4,

$$-v\frac{\partial n_3}{\partial \xi} + \frac{\partial n_1}{\partial \tau} + \frac{\partial}{\partial \xi}(n_1 u_2 + n_2 u_1 + u_3) = 0, \qquad (6.8)$$

$$-v\frac{\partial u_3}{\partial \xi} + \frac{\partial u_1}{\partial \tau} + \frac{\partial}{\partial \xi}(u_1 u_2) + \frac{\partial \phi_3}{\partial \xi} = 0, \qquad (6.9)$$

$$\frac{\partial^2 \phi_1}{\partial \xi^2} + n_3 - a(f\alpha_c + (1-f)\alpha_h)\phi_3 - 2b(f\alpha_c^2 + (1-f)\alpha_h^2)(\phi_1\phi_2)$$
$$-c(f\alpha_c^3 + (1-f)\alpha_h^3)\phi_1^3 = 0. \qquad (6.10)$$

Differentiating equation (6.10) with respect to ξ and removing all higher order perturbed terms using equations (6.6)-(6.8), the modified KdV (mKdV) equation is obtained as

$$\frac{\partial \phi_1}{\partial \tau} + C\phi_1^2 \frac{\partial \phi_1}{\partial \xi} + B\frac{\partial^3 \phi_1}{\partial \xi^3} = 0, \tag{6.11}$$

where $C = \frac{v^3}{2}(\frac{3}{v^6} + \frac{3}{v^2}b(f\alpha_c^2 + (1-f)\alpha_h^2) - 3c(f\alpha_c^3 + (1-f)\alpha_h^3))$ and
$c = \dfrac{(\kappa^2 - \frac{1}{4})(\kappa + \frac{3}{2})}{6(\kappa - \frac{3}{2})^3}.$

6.2.3 Formation of dynamical system and phase portraits

Using the wave frame $\phi_1(\xi, \tau) = \Phi(\chi)$ with $\chi = \xi - V\tau$ in equation (6.11), one can obtain

$$\begin{cases} \frac{d\Phi}{d\chi} = z, \\ \frac{dz}{d\chi} = \frac{1}{B}\left(V\Phi - \frac{C}{3}\Phi^3\right), \end{cases} \tag{6.12}$$

where V is velocity of the wave profile. Now, phase portrait profiles of the system (6.12) for the mKdV equation (6.11) will be studied with some special values of κ, α_h, α_c, f and V. For that purpose, fixed points of the system (6.12) are obtained by solving the following equations simultaneously

$$\frac{d\Phi}{d\chi} = 0 \text{ and } \frac{dz}{d\chi} = 0,$$

which imply

$$z = 0 \text{ and } \Phi\frac{1}{B}(V - \frac{C}{3}\Phi^2) = 0,$$

$$\Rightarrow z = 0 \text{ and } \Phi = 0, \ \pm\sqrt{\frac{3V}{C}}.$$

Therefore, there are three fixed points $E_0(\Phi_0, 0)$, $E_1(\Phi_1, 0)$ and $E_2(\Phi_2, 0)$ of the system (6.12), where

$$\Phi_0 = 0, \ \Phi_1 = \sqrt{\frac{3V}{C}}, \text{ and } \Phi_2 = -\sqrt{\frac{3V}{C}}.$$

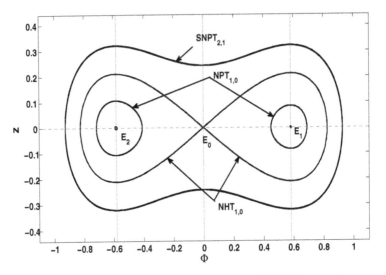

Figure 6.6: Phase portrait of dynamical system (6.12) for $\kappa = 2$, $\alpha_h = 0.32$, $\alpha_c = 1.1$, $f = 0.1$ and $V = 0.1$.

The Jacobian matrix J for the system (6.12) is

$$J = \begin{pmatrix} 0 & 1 \\ \frac{1}{B}(V - C\Phi_i^2) & 0 \end{pmatrix},$$

and determinant of J is expressed by

$$M = det J(\Phi_i, 0) = -\frac{1}{B}(V - C\Phi_i^2),$$

where $i = 0, 1, 2$. If $M < 0$, then fixed point $E_i(\Phi_i, 0)$ is a saddle node and for $M > 0$, fixed point $E_i(\Phi_i, 0)$ is a center [3]. Applying the phase plane analysis of dynamical systems [3], phase portrait profile of the system (6.12) is presented in Figure 6.6.

Through computation, phase portrait profile of equation (6.12) for the mKdV equation (6.11) relying on system parameters κ, α_h, α_c, f and V is presented in Figure 6.6. A couple of nonlinear homoclinic trajectories (NHT$_{1,0}$) enclosing one fixed point on both sides with no separatrix layer is presented. Any qualitative trajectory in the phase profile corresponds to a traveling wave solution. Here, NHT$_{1,0}$ enclosing fixed point E_1 corresponds to compressive ion-acoustic solitary wave (IASW). Similarly, NHT$_{1,0}$ enclosing fixed point E_2 corresponds to rarefactive IASW. Nonlinear periodic trajectories (NPT$_{1,0}$)

around E_1 and E_2 correspond to nonlinear periodic wave (NPIAW) solutions. There also exists a class of supernonlinear periodic trajectory (SNPT$_{2,1}$) enclosing fixed points E_0, E_1 and E_2 with one separatrix layer. Here, SNPT$_{2,1}$ corresponds to supernonlinear periodic ion-acoustic wave (SNPIAW) solution of the mKdV equation (6.11). It is observed numerically that small-amplitude SNPIAW features for the mKdV equation (6.11) exist in the considered plasma system.

6.2.4 Wave solutions

In Figure 6.7, we present the effect of α_h on small-amplitude SNPIAW solution of the mKdV equation (6.11) with system parametric data same as Figure 6.6. We observe from Figure 6.7 that, when the temperature of hot electrons is lowered, amplitude and width of SNPIWs are diminished.

SNPT$_{2,1}$ presented in Figure 6.6 corresponds to SNPIAW solution which is shown by Figure 6.8. Effect of κ on SNPIAW solution of the mKdV equation (6.11) is shown in Figure 6.8 with parametric data same as Figure 6.6. One can observe that amplitude and width of SNPIAWs are flourished when spectral index (κ) of electrons is increased.

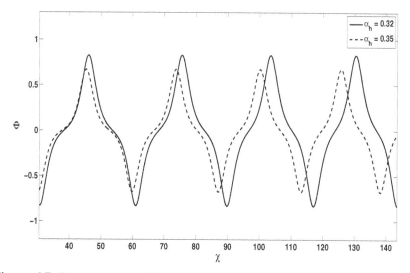

Figure 6.7: Phase portrait of dynamical system (6.12) for $\kappa = 2, \alpha_c = 1.1$, $f = 0.1$ and $V = 0.1$.

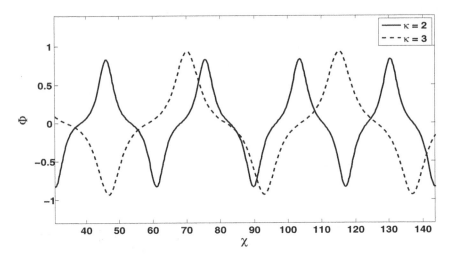

Figure 6.8: Phase portrait of dynamical system (6.12) for $\alpha_h = 0.32, \alpha_c = 1.1, f = 0.1$ and $V = 0.1$.

6.3 Bifurcation of supernonlinear dust-acoustic waves

Dust-acoustic wave is one kind of mode [4]-[6] in dusty plasmas due to the presence of additional micro-sized charged components [7]. A new type of wave characterized by nontrivial topology of the phase portrait in plasmas composed of multi-components is known as supernonlinear wave [1]. Very recently, the dynamical features of ion-acoustic superperiodic waves were studied by Saha et al. [8] by employing the theory of phase plane analysis [9]-[10]. Here, supernonlinear dust-acoustic waves are investigated in a multi-component dusty plasma with negatively charged dust particles, Maxwellian electrons and q-nonextensive ions employing phase plane analysis.

6.3.1 Basic equations

The normalized basic equations of an unmagnetized dusty plasma are composed of mobile dust particles with negative charge, Maxwellian distributed electrons and q-nonextensive distributed ions as

$$\frac{\partial n}{\partial t} + \frac{\partial (nu)}{\partial x} = 0, \tag{6.13}$$

$$\frac{\partial u}{\partial t} + u\frac{\partial u}{\partial x} = \frac{\partial \phi}{\partial x}, \tag{6.14}$$

$$\frac{\partial^2 \phi}{\partial x^2} = n_e + n - n_i, \tag{6.15}$$

$$n_e = \mu_1 \exp(\sigma\phi), \tag{6.16}$$

$$n_i = \mu_2 [1 - (q-1)\phi]^{\frac{1}{q-1}+\frac{1}{2}}, \tag{6.17}$$

where $\mu_1 = \dfrac{n_{e0}}{n_0}$, $\mu_2 = \dfrac{n_{i0}}{n_0}$ and $\sigma = \dfrac{T_i}{T_e}$. In this case, (n_i, n, n_e) are number densities of (ions, dusts, electrons), respectively. One can normalize velocity u of dusts by $C_s = \left(\dfrac{k_B T_e}{m_d}\right)^{\frac{1}{2}}$, electrostatic potential ϕ by $\dfrac{k_B T_d}{e}$, time t by $\omega_d^{-1} = \left(\dfrac{m_d}{4\pi n_0 e^2}\right)^{\frac{1}{2}}$, space variable x by the Debye length $\lambda_D = \left(\dfrac{k_B T_e}{4\pi n_0 e^2}\right)^{\frac{1}{2}}$, (n, n_e, n_i) by their corresponding unperturbed values (n_0, n_{e0}, n_{i0}), where q, k_B, m_d, ω_d and e are nonextensive parameter, Boltzmann constant, mass of dusts, dusty plasma frequency, magnitude of electron charge, respectively.

6.3.2 *Formation of dynamical system and phase portraits*

Let us consider a traveling wave frame with velocity M as

$$\xi = x - Mt, \tag{6.18}$$

where M is also known as the Mach number. Using equation (6.18) and boundary conditions $u \to 0$, $n \to 1$ and $\phi \to 0$ as $\xi \to \pm\infty$, in equations (6.13) and (6.14), one can obtain

$$n = \frac{M}{\sqrt{M^2 - 2\phi}}. \tag{6.19}$$

Using equations (6.16)-(6.19) in equation (6.15), we get

$$\frac{d^2\phi}{d\xi^2} = a\phi + b\phi^2 + c\phi^3 + d\phi^4, \tag{6.20}$$

where

$$a = -\frac{1}{M^2} + \sigma\mu_1 + \frac{(q+1)}{2}\mu_2,$$

$$b = \frac{3}{2M^4} + \frac{1}{2}\sigma^2\mu_1 + \frac{(q+1)(q-3)}{8}\mu_2,$$

$$c = -\frac{5}{2M^6} + \frac{1}{6}\sigma^3\mu_1 + \frac{(q+1)(q-3)(3q-5)}{48}\mu_2,$$

and

$$d = \frac{35}{8M^8} + \frac{1}{24}\sigma^4\mu_1 + \frac{(q+1)(q-3)(3q-5)(5q-7)}{384}\mu_2.$$

The system (6.20) can be expressed as the following coupled system

$$\begin{cases} \frac{d\phi}{d\zeta} = z, \\ \frac{dz}{d\zeta} = a\phi + b\phi^2 + c\phi^3 + d\phi^4. \end{cases} \tag{6.21}$$

The corresponding Hamiltonian function $H(\phi, z)$ for the coupled system (6.21) is obtained as

$$H(\phi, z) = \frac{z^2}{2} - a\frac{\phi^2}{2} - b\frac{\phi^3}{3} - c\frac{\phi^4}{4} - d\frac{\phi^5}{5} = h. \tag{6.22}$$

The relation $h_i = H(\phi, z)$ provides a trajectory in the (ϕ, z)-phase plane passing through a point (ϕ_i, z_i), where $h_i = H(\phi_i, z_i)$. Using equation (6.22) in equation (6.21), we present all phase plots and corresponding wave solutions in Figures 6.9-6.17.

6.3.3 Wave solutions

One can study supernonlinear dust-acoustic wave solutions based on the coupled system (6.21) using phase plane analysis varying the physical parameters q, σ, μ_1, μ_2, and M. Phase plots of the coupled system (6.21) can vary significantly based on number of fixed points with their stabilities and separatrix layers [1]. Any trajectory in a phase plot of the coupled system (6.21) refers to a dust-acoustic wave solution of the dusty plasma system. One can denote $NHT_{1,0}$ for nonlinear homoclinic trajectory, $NPT_{1,0}$ for nonlinear periodic trajectory, $SNPT_{2,1}$ for supernonlinear periodic trajectory and $SNHT_{2,1}$ for supernonlinear homoclinic trajectory. There is a one to one correspondence between trajectories in the phase plot and dust-acoustic wave solutions of the dusty plasma. Hence, to obtain all possible supernonlinear dust-acoustic wave solutions of the dusty plasma system, one needs to find all possible supernonlinear trajectories of the coupled system (6.21) varying suitable physical parameters.

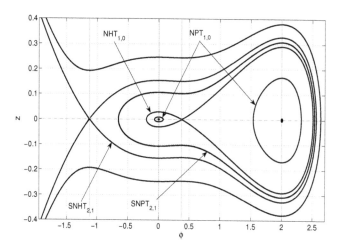

Figure 6.9: Phase plot of the system (6.21) for $q = -0.01$, $\mu_1 = 0.14$, $\mu_2 = 0.09$, $\sigma = 0.7$ and $M = 2.4$.

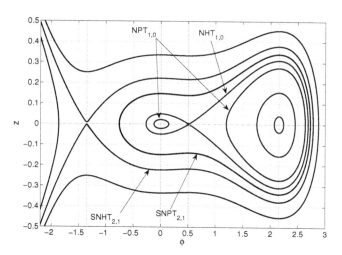

Figure 6.10: Phase plot of the system for $q = 0.1$, $\mu_1 = 0.14$, $\mu_2 = 0.09$, $\sigma = 0.7$ and $M = 2.4$.

For all fixed points of the coupled system (6.21), the following equations $\frac{d\phi}{d\xi} = 0$ and $\frac{dz}{d\xi} = 0$ are solved simultaneously. Then, one can find

$$z = 0 \text{ and } \phi(\phi^3 + p\phi^2 + q\phi + r) = 0, \tag{6.23}$$

with $p = \frac{c}{d}$, $q = \frac{b}{d}$ and $r = \frac{a}{d}$.

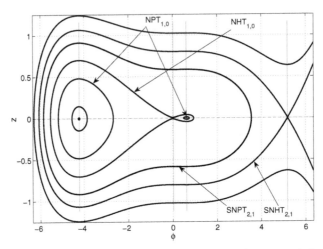

Figure 6.11: Phase plot of the system (6.21) for $q = 1.6$, $\mu_1 = 0.01$, $\mu_2 = 0.09$, $\sigma = 0.7$ and $M = 3$.

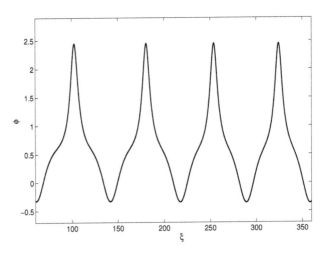

Figure 6.12: Supernonlinear periodic wave solution for $q = -0.01$, $\mu_1 = 0.14$, $\mu_2 = 0.09$, $\sigma = 0.7$ and $M = 2.4$.

When $\frac{h^2}{4} + \frac{g^3}{27} > 0$, the coupled system (6.21) has two fixed points at $F_0(\phi_0, 0)$ and $F_1(\phi_1, 0)$, where $\phi_0 = 0$, and $\phi_1 = A + B$, with $A = \sqrt[3]{-\frac{h}{2} + \sqrt{\frac{h^2}{4} + \frac{g^3}{27}}}$, $B = \sqrt[3]{-\frac{h}{2} - \sqrt{\frac{h^2}{4} + \frac{g^3}{27}}}$, $g = \frac{1}{3}(3q - p^2)$ and $h = \frac{1}{27}(2p^3 - 9pq + 27r)$.

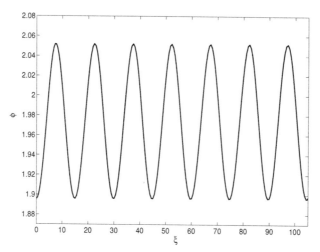

Figure 6.13: Nonlinear periodic wave solution for $q = -0.01$, $\mu_1 = 0.14$, $\mu_2 = 0.09$, $\sigma = 0.7$ and $M = 2.4$.

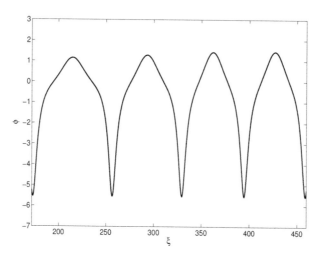

Figure 6.14: Supernonlinear periodic wave solution for $q = 0.1$, $\mu_1 = 0.14$, $\mu_2 = 0.09$, $\sigma = 0.7$ and $M = 2.4$.

When $\frac{h^2}{4} + \frac{g^3}{27} < 0$, then the coupled system (6.21) has four fixed points at $F_0(\phi_0, 0)$, $F_2(\phi_2, 0)$, $F_3(\phi_3, 0)$ and $F_4(\phi_4, 0)$, with $\phi_0 = 0$, and $\phi_{2,3,4} =$

$$-\frac{p}{3} + 2\sqrt{-\frac{g}{3}}\cos\left(\frac{\psi}{3} + \frac{2k\pi}{3}\right), \quad k = 0,1,2, \cos\psi = \begin{cases} \sqrt{\dfrac{h^2/4}{-g^3/27}}, & \text{if } h < 0; \\[3mm] -\sqrt{\dfrac{h^2/4}{-g^3/27}}, & \text{if } h > 0. \end{cases}$$

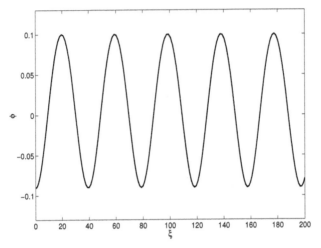

Figure 6.15: Nonlinear periodic wave solution for $q = 0.1$, $\mu_1 = 0.14$, $\mu_2 = 0.09$, $\sigma = 0.7$ and $M = 2.4$.

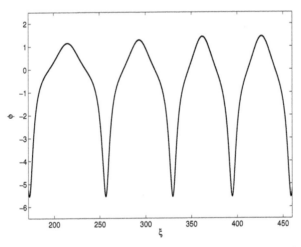

Figure 6.16: Supernonlinear periodic wave solution for $q = 1.6$, $\mu_1 = 0.01$, $\mu_2 = 0.09$, $\sigma = 0.7$ and $M = 3$.

If J is the determinant value of the Jacobian matrix at $(\phi_i, 0)$ of the coupled system (6.21), then one can obtain

$$J(\phi_i, 0) = -(a + 2b\phi_i + 3c\phi_i^2 + 4d\phi_i^3). \tag{6.24}$$

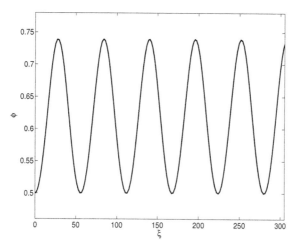

Figure 6.17: Nonlinear periodic wave solution for $q = 1.6$, $\mu_1 = 0.01$, $\mu_2 = 0.09$, $\sigma = 0.7$ and $M = 3$.

Considering suitable values of physical parameters q, μ_1, μ_2, σ and M, all probable phase plots of the coupled system (6.21) and corresponding periodic and superperiodic wave solutions are depicted in Figures 6.9-6.17.

The plasma system supports supernonlinear dust-acoustic periodic waves (Figures 6.12, 6.14 and 6.16) corresponding to the $\text{SNPT}_{2,1}$ (Figures 6.9, 6.10 and 6.11). Furthermore, the system also supports nonlinear periodic waves (Figures 6.13, 6.15 and 6.17). It is observed that q plays a crucial role in the stability of the equilibrium point of the system (6.21) in the ranges $-1 < q < 1$ and $q > 1$.

6.4 Bifurcation of supernonlinear electron-acoustic waves (EAWs)

The basics of EAWs were proposed by Fried and Gould [11]. The authors observed that besides dust-acoustic, ion-acoustic and dust-ion-acoustic waves, another highly energetic acoustic wave can exist in plasmas which is effectively different from the others [11, 12, 13]. This kind of electrostatic wave can move in magnetized [14, 15] as well as unmagnetized plasma [16, 17]. In general, the EAWs in plasmas involving two distinct temperatures for electrons, i.e., cold electrons and hot electrons. In this case, dynamics of cold electrons are considered and ions remain motionless to create a neutralized background, where hot electrons are responsible for building the restoring force for the nonlinear EAWs.

6.4.1 Basic equations

One can consider the simple case of one dimensional EAWs in a homogeneous plasma composed of mobile cold electrons, immobile ions, and q-nonextensive distributed hot electrons. The basic equations [18] for the dynamics of EAWs are:

$$\frac{\partial n}{\partial t} + \frac{\partial}{\partial x}(nu) = 0, \tag{6.25}$$

$$\frac{\partial u}{\partial t} + u\frac{\partial u}{\partial x} = \alpha\frac{\partial \phi}{\partial x}, \tag{6.26}$$

$$\frac{\partial^2 \phi}{\partial x^2} = n_e + \frac{1}{\alpha}n - (1 + \frac{1}{\alpha}), \tag{6.27}$$

where $\alpha = \frac{n_{e0}}{n_0} > 1$. Here, n (n_e) denotes number density of cold (hot) electrons. Here, ϕ denotes electrostatic potential, u denotes velocity of cold electrons, T_h is temperature of hot electrons, m_h denotes electron mass, e is charge of electrons, and k_B is the Boltzmann constant. One can refer to space variable and time with x and t, respectively. One can normalize n by n_0, u by $C_s = (k_B T_h/\alpha m_e)^{\frac{1}{2}}$, ϕ by $k_B T_h/e$, t by $\omega_{pc}^{-1} = (m_e/4\pi n_0 e^2)^{\frac{1}{2}}$ and x by $\lambda_{D-} = (k_B T_h/4\pi e^2 n_{e0})^{\frac{1}{2}}$.

The velocity distribution for electrons is given by

$$f_e(v) = C_q\{1 + (q-1)[\frac{m_e v^2}{2k_B T_e} - \frac{e\phi}{k_B T_e}]\}^{\frac{1}{(q-1)}},$$

where q is a nonextensive parameter and T_e is temperature of electrons. The normalization constant C_q is given by

$$C_q = n_{e0}\frac{\Gamma(\frac{1}{1-q})}{\Gamma(\frac{1}{1-q}-\frac{1}{2})}\sqrt{\frac{m_e(1-q)}{2\pi k_B T_e}} \quad \text{for } -1 < q < 1,$$

and

$$C_q = n_{e0}\frac{1+q}{2}\frac{\Gamma(\frac{1}{q-1}+\frac{1}{2})}{\Gamma(\frac{1}{q-1})}\sqrt{\frac{m_e(q-1)}{2\pi k_B T_e}} \quad \text{for } q > 1.$$

Integrating $f_e(v)$ over all velocity spaces, one can obtain

$$n_e = n_{e0}\{1 + (q-1)\tfrac{e\phi}{k_B T_e}\}^{1/(q-1)+1/2}.$$

The normalized hot electrons [18] number density is given by

$$n_e = \{1 + (q-1)\phi\}^{\frac{1}{q-1}+\frac{1}{2}}. \tag{6.28}$$

6.4.2 The evolution equation and dynamical system

To derive the evolution equation "nonlinear Schrodinger equation (NSE)" using RPT, the independent variables are stretched as

$$\xi = \varepsilon(x - Vt), \quad \tau = \varepsilon^2 t, \tag{6.29}$$

and dependent variables are expanded as

$$\begin{cases} n = 1 + \sum_{m=1}^{\infty} \varepsilon^m \sum_{l=-m}^{m} n_l^{(m)}(\xi, \tau)\exp[il(kx - \omega t)], \\ u = \sum_{m=1}^{\infty} \varepsilon^m \sum_{l=-m}^{m} u_l^{(m)}(\xi, \tau)\exp[il(kx - \omega t)], \\ \phi = \sum_{m=1}^{\infty} \varepsilon^m \sum_{l=-m}^{m} \phi_l^{(m)}(\xi, \tau)\exp[il(kx - \omega t)], \end{cases} \tag{6.30}$$

where V is the group velocity and ε is a small parameter $(0 < \varepsilon << 1)$. Using equations (6.28)-(6.30) in equations (6.25)-(6.27), one can obtain the NSE equation [19] as

$$i\frac{\partial \Phi}{\partial \Phi} + P\frac{\partial^2 \Phi}{\partial \xi^2} + Q|\Phi^2|\eta = 0, \tag{6.31}$$

where $\Phi = \Phi_1^{(1)}$, $P = -3C_1\dfrac{\omega^5}{2k^4}$ and $Q = \dfrac{1}{12k^4\omega^5\beta_1}[18C_3k^4\omega^8\beta_1 + 4C_2\omega^8\beta_2 + 3k^8(2\beta^3 - \alpha^2\beta^4) + 12C_2k^4\beta^5\omega^4]$,

with $\omega = \dfrac{k}{\sqrt{C_1 + k^2}}$, $C_1 = \frac{q+1}{2}$, $C_2 = \frac{(q+1)(3-q)}{8}$, $C_3 = \dfrac{(1+q)(3-q)(5-3q)}{48}$,

$\beta_1 = (3 - 3\omega^2 + \omega^4)$, $\beta_2 = (3 - 9\omega^2 + 5\omega^4)$, $\beta_3 = (9 - 9\omega^2 + 4\omega^4)$, $\beta_4 = (9 + 6\omega^2 - 12\omega^4 + 5\omega^6)$ and $\beta_5 = (-6 - \alpha\beta_1 + 4\omega^2)$.

For the study of nonlinear and supernonlinear EAWs, one can take $\chi = l\xi - v\tau$ and $\eta(\chi) = \Psi(\chi)\exp(i\beta\chi)$ in the NLS equation (6.31) to find an ordinary differential equation as

$$Pl^2\frac{d^2\Psi}{d\chi^2} + (\beta v\Psi - \beta^2 Pl^2\Psi + Q|\Psi^2|\Psi) + i(-v + 2Pl^2\beta)\frac{d\Psi}{d\chi} = 0, \tag{6.32}$$

where v denotes wave velocity and $0 < l < 1$. For the further study, one can consider the real part of equation (6.32) and acquire

$$\frac{d^2\Psi}{d\chi^2} = \left(\beta^2 - \frac{\beta v}{Pl^2}\right)\Psi - \frac{Q}{Pl^2}\Psi^3. \tag{6.33}$$

The system (6.33) can be expressed as a coupled system

$$\begin{cases} \frac{d\Psi}{d\chi} = y, \\ \frac{dy}{d\chi} = M_1\Psi - M_2\Psi^3, \end{cases} \tag{6.34}$$

where $M_1 = \beta^2 - \frac{\beta v}{Pl^2}$ and $M_2 = \frac{Q}{Pl^2}$.

The corresponding Hamiltonian function $H(\Psi, y)$ for the coupled system (6.34) is obtained as

$$H(\Psi, y) = \frac{y^2}{2} - \frac{M_1\Psi^2}{2} + \frac{M_2\Psi^4}{4} = h, \tag{6.35}$$

and we obtain

$$\frac{d\Psi}{d\chi} = \sqrt{\frac{-M_2}{2}}\sqrt{(p_1 - \Psi)(\Psi - p_2)(\Psi - p_3)(\Psi - p_4)}, \tag{6.36}$$

where equation $h_i = \frac{M_2}{2}\left(\Psi^4 - \frac{2M_1}{M_2}\Psi^2\right)$ gives four roots p_1, p_2, p_3 and p_4 with $h_i = H(\Psi_i, y_i)$ and (Ψ_i, y_i) is any point on the (Ψ, y)-plane. Equation (6.36) acquires the wave solution of equation (6.31) for any trajectory passing through (Ψ_i, y_i).

The coupled system (6.34) has three fixed (equilibrium) points F_i at $(\Psi_i, 0)$, for $i = 0, 1, 2$. Based on $H(\Psi, y) = 0$ at F_0, one can present homoclinic trajectories enclosing fixed points F_1 and F_2 in Figure 6.18. Again, depending on $H(\Psi, y) = h_i$ around F_1 and F_2, two families of nonlinear periodic trajectories exist. A family of periodic trajectories enclosing the fixed points F_0, F_1 and F_2 with one separatrix is presented. These periodic trajectories are called supernonlinear periodic trajectories and are responsible for supernonlinear periodic waves.

The general ranges for q are known as $-1 < q < 1$ and $q > 1$. However, based on current studies [20, 21], the appropriate range for q is $1/3 < q < 1$ due to

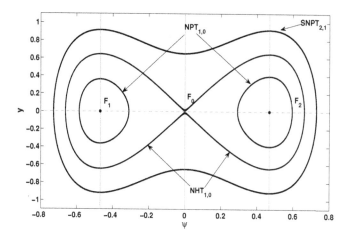

Figure 6.18: Phase portrait for $q = 0.8$ with $\alpha = 1.5$, $\beta = 0.2$, $l = 0.3$, $k = 0.3$ and $v = 0.7$.

the physical situation of energy equipartition. So, a suitable range for q is $1/3 < q < 1$.

The Figure 6.18 shows the phase space of the coupled system (6.34) considering $q = 0.5$, $\alpha = 1.5$, $\beta = 0.2$, $l = 0.3$, $k = 0.3$ and $v = 0.2$. Here, $NHT_{1,0}$ denotes nonlinear homoclinic trajectory, $NPT_{1,0}$ denotes nonlinear periodic trajectory and $SNPT_{2,1}$ denotes superperiodic trajectory.

6.4.3 Wave solutions

In Figures 6.19-6.22, we present various EAWs for different trajectories in the phase space given in Figure 6.18. We show periodic wave solutions in Figures 6.19, 6.20, and superperiodic wave solutions in Figures 6.21, 6.22, depending on q and v. One can observe from Figures 6.19 and 6.21 that for suitable higher values of q with fixed values of other parameters, amplitude of periodic wave solutions increases, but width decreases. On the other hand, amplitude and width of superperiodic EAWs diminish. If velocity (v) of the wave is increased, then amplitude of EA periodic wave decays and width grows. However, in the case of superperiodic EAWs, amplitude increases and its width decrease in Figures 6.20 and 6.22 when velocity v of the wave increases.

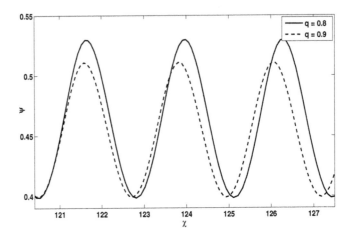

Figure 6.19: Nonlinear periodic solutions for variation of q with other parameters same as Figure 6.18.

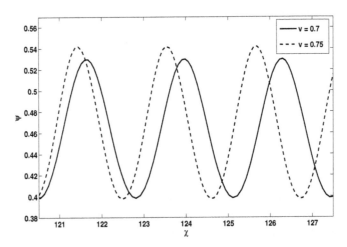

Figure 6.20: Nonlinear periodic solutions for variation of v with other parameters same as Figure 6.18.

The electron-acoustic solitary wave solutions of equation (6.31) with compressive and rarefactive types are as follows

$$\Psi = \pm\sqrt{\frac{2M_1}{M_2}}\,sech(\sqrt{M_1}\chi). \tag{6.37}$$

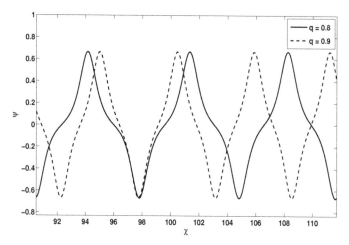

Figure 6.21: Supernonlinear periodic solutions for variation of q with other parameters same as Figure 6.18.

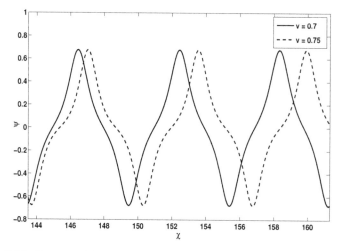

Figure 6.22: Supernonlinear periodic solutions for variation of v with other parameters same as Figure 6.18.

Here, negative (positive) sign in equation (6.37) refers to rarefactive (compressive) EAWs.

Effects of the parameters q and v are shown on electron-acoustic compressive and rarefactive solitary waves in Figures 6.23-6.26. It is seen that EASW

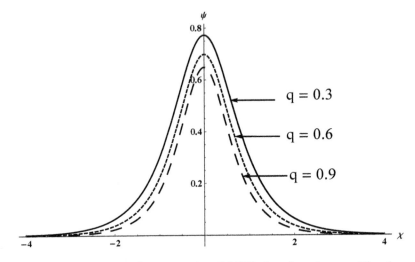

Figure 6.23: Variations of compressive EASWs by changing q with other parameters same as Figure 6.18.

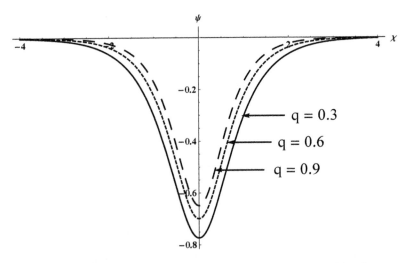

Figure 6.24: Variations of rarefactive EASWs by changing q with other parameters same as Figure 6.18.

solutions flourish as q decreases. Furthermore, EASWs become spiky when velocity v of the wave profile is enhanced. It is seen that q-nonextensive parameter and v significantly affect nonlinear EAWs as well as supernonlinear EAWs.

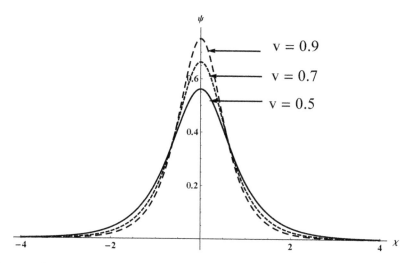

Figure 6.25: Variations of compressive EASWs by changing v with other parameters same as Figure 6.18.

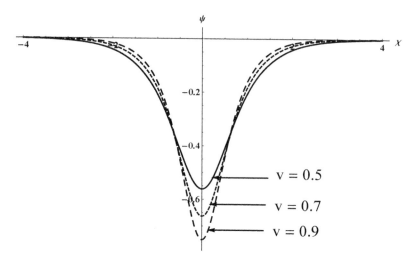

Figure 6.26: Variations of rarefactive EASWs by changing v with other parameters same as Figure 6.18.

References

[1] A. E. Dubinov and D. Yu. Kolotkov, IEEE Transactions on Plasma Science, 40(5): 1429–1433 (2012).

[2] D. P. Chapagai, J. Tamang and A. Saha, Zeitschrift für Naturforschung A, 7593: 183–191 (2020).

[3] S. H. Strogatz, Nonlinear Dynamics and Chaos, Westview Press (USA) (2007).

[4] N. N. Rao, M. Y. Yu and P. K. Shukla, Planet Space Sci., 38: 543 (1990).

[5] I. Kourakis and P. K. Shukla, Eur. Phys. J. D., 30: 97 (2004).

[6] F. Melandso, Phys. Plasmas, 3: 3890 (1996).

[7] P. K. Shukla and A. A. Mamun, Plasma Physics and Controlled Fusion, 44 (2002).

[8] A. Saha and J. Tamang, Advances in Space Research, 63(5): 1596–1606 (2019).

[9] J. Guckenheimer and P. J. Holmes, Nonlinear Oscillations Dynamical Systems and Bifurcations of Vector Fields. New York: Springer-Verlag (1983).

[10] M. Lakshmanan and S. Rajasekar, Nonlinear Dynamics: Integrability, Chaos and Patterns, Springer-Verlag (2003).

[11] B. D. Fried and R. W. Gould, Phys. Fluids, 4: 139 (1961).

[12] H. Demiray, Phys. Plasmas, 23: 032109 (2016).

[13] K. Watanabe and T. Taniuti, 43: 1819 (1977).

[14] P. Shukla, A. Mamun and B. Eliasson, Geophys. Res. Lett., 31: L07803 (2004).

[15] M. Shalaby, S. El-Labany, R. Sabry and L. El-Sherif, Phys. Plasmas, 18: 062305 (2011) .

[16] W. El-Taibany and W. M. Moslem, Phys. Plasmas, 12: 032307 (2005).

[17] P. Eslami, M. Mottaghizadeh and H. R. Pakzad, Phys. Plasmas, 18: 102313 (2011).

[18] A. Abdikian, J. Tamang and A. Saha, Commun. Theor. Phys., 72: 075502 (2020).

[19] A. S. Bains, M. Tribeche and T. S. Gill, Phys. Plasmas, 18: 022108 (2011).

[20] L. Ait Gougam and M. Tribeche, Physica A, 407: 226 (2014).

[21] S. A. El-Tantawy, E. I. El-Awady and M. Tribeche, Phys. Plasmas, 22: 113705 (2015).

Chapter 7

Chaos, Multistability and Stable Oscillation in Plasmas

7.1 Chaos in a conservative dusty plasma

Over the last few years, the study of dusty plasmas has been significant as it contributes widely in various research areas. Dusty plasmas are studied not only for the educational perspective, but also for their wide implications found in modern astrophysics and space, fusion devices, semiconductor technology, biophysics and plasma chemistry [1]-[2]. For the first time, the presence of low frequency dust-ion-acoustic waves (DIAWs) was reported by Shukla and Silin [3] in a dusty plasma containing inertial ions with Maxwellian electrons, and immobile negatively charged dust particles. Based on various kinds of dust charged particles in dusty plasmas, different classes of new wave modes are proposed, such as, DIA mode, dust drift mode, dust acoustic mode, Shukla-Varma mode, dust lattice mode, and dust Berstain-Green-Kruskal mode [4]-[8]. The dust-ion-acoustic solitary waves (DIASWs) were examined experimentally and theoretically by numerous researchers [9]-[17]. By considering a dusty double-plasma device, the experimental observations of ion-acoustic solitary waves were performed by Nakamura and Sarma [18] in a dusty plasma. They reported that the channel of oscillation became insubstantial when dust grains were added into the plasma. Later, the fundamental characteristics of DIASWs were investigated by Anowar and Mamun [19] in a hot adiabatic magnetized dusty plasma.

The deviation of particles from the classical Maxwellian limit has been observed for the particles possessing high energy tails [20]-[21]. This property exists in particles with the long range interactions caused by interactions between external force fields and plasmas [20]-[21]. Such particles are significant to characterize the non-equilibrium stationary states, such as, global correlations [22] in the self-gravitating systems, long-range interactions [23] in the plasma particles, long-time memory effects or in a multi-fractal space. In such cases, plasma components obey non-Maxwellian flow that is a particular class of Tsallis's velocity distribution [24]. This Tsallis's distribution [24] is also defined as q-distribution, where parameter q is known to measure the nonextensivity of plasma particles. The concept of nonextensive entropy proposed by Tsallis [24] had an influential implication on the theory proposed by Renyi [25]. Thus, the studies of plasmas with nonextensive distribution have grabbed a great attention in different plasma systems [26]-[28].

It is noteworthy that the effects of external periodic perturbations existing in plasmas cause interruption on the integrability of a system [29]-[31]. Such external periodic perturbation may change based on various physical conditions. Recently, the investigation of systems with external periodic forcing has gained remarkable recognition. The existence of quasiperiodic and chaotic features cannot be depicted for a nonlinear system that is completely integrable. However, the presence of an external periodic perturbation may indicate quasiperiodic or chaotic motions existing in a system. Applying an external periodic perturbation to a system, the quasiperiodic behavior of nonlinear waves in quantum plasmas was investigated by Sahu et al. [32] for the first time in the literature. Later, they also reported the quasiperiodic behavior in quantum plasmas due to the occurrence of Bohm potential [33]. The dynamical feature of nonlinear waves in quantum magnetoplasma was studied by Zhen et al. [34] and later, they also investigated solitary solution and chaotic feature under the extended ZK equation in magnetized dusty plasma [35]. This gave a motivation to investigate quasiperiodic and chaotic features [36]-[39] of DIAWs in unmagnetized plasmas.

7.1.1 Basic equations

Consider a three-constituent dusty plasma composed of cold inertial ions, non-inertial q-nonextensive electrons and immobile negatively charged dusts. At unperturbed state, $n_{e0} = n_0 - Z_d n_{d0}$ is maintained, with Z_d being the number of electrons residing onto the dust particle. Here, n_0, n_{e0} and n_{d0} denote ion, electron and dust number densities, respectively. In this case, phase speed is higher than thermal speed of ions and smaller than that of electrons. The motions of DIAWs [40] are described by the following equations:

$$\frac{\partial n}{\partial t} + \frac{\partial (nu)}{\partial x} = 0, \tag{7.1}$$

$$\frac{\partial u}{\partial t} + u\frac{\partial u}{\partial x} = -\frac{\partial \phi}{\partial x}, \tag{7.2}$$

$$\frac{\partial^2 \phi}{\partial x^2} = (1-\mu)n_e - n + \mu, \tag{7.3}$$

where $\mu = \dfrac{Z_d n_{d0}}{n_0}$.

The number density of normalized electrons [41] is of the form

$$n_e = \{1 + (q-1)\phi\}^{\frac{1}{q-1}+\frac{1}{2}}, \tag{7.4}$$

where q is called nonextensive parameter and it can take values which are greater than -1.

7.1.2 Multiperiodic, quasiperiodic and chaotic oscillations

Multiperiodic, quasiperiodic and chaotic features of DIAWs are studied. For this purpose, one traveling wave transfiguration $\xi = x - vt$ is assumed with wave velocity v. Applying ξ and boundary conditions $n = 1, u = 0$ and $\phi = 0$ as $\xi \to \pm\infty$ in equations (7.1) and (7.2), one can get

$$n = \frac{v}{\sqrt{v^2 - 2\phi}}. \tag{7.5}$$

Considering up to 3rd degree terms, from equations (7.3)-(7.5) one can obtain

$$\frac{d^2\phi}{d\xi^2} = a\phi + b\phi^2 + c\phi^3, \tag{7.6}$$

here $a = \dfrac{(1-\mu)(1+q)}{2} - \dfrac{1}{v^2}$, $b = \dfrac{(1-\mu)(1+q)(3-q)}{8} - \dfrac{3}{2v^4}$, and $c = \dfrac{(1-\mu)(1+q)(3-q)(5-3q)}{48} - \dfrac{5}{2v^4}$.

The following dynamical system can be formed from equation (7.6)

$$\begin{cases} \frac{d\phi}{d\xi} = z, \\ \frac{dz}{d\xi} = a\phi + b\phi^2 + c\phi^3. \end{cases} \tag{7.7}$$

Considering an external forcing $f_0 \cos(\omega\xi)$ to the system (7.7), one can obtain

$$\begin{cases} \frac{d\phi}{d\xi} = z, \\ \frac{dz}{d\xi} = a\phi + b\phi^2 + c\phi^3 + f_0 \cos(\omega\xi), \end{cases} \tag{7.8}$$

where f_0 is strength and ω is the frequency of the external forcing. The system (7.8) is a nonautonomous dynamical system.

The system (7.8) is rewritten in the form of an autonomous dynamical system (ADS) as

$$\begin{cases} \frac{d\phi}{d\xi} = z, \\ \frac{dz}{d\xi} = a\phi + b\phi^2 + c\phi^3 + f_0 \cos(w), \\ \frac{dw}{d\xi} = \omega. \end{cases} \tag{7.9}$$

Phase plot of the ADS (7.9) is presented for $q = 2, \mu = 0.54, v = 0.8, f_0 = 0.9$ and $\omega = 1.9$ with initial condition (-0.2309, 0.14602, 0) in Figure 7.1 and the corresponding time series plot for ϕ is presented in Figure 7.2 with fixed parametric values and initial condition, as in Figure 7.1. In this case, multi-periodic oscillation of the DIAWs is observed.

Phase plot of the ADS (7.9) is shown for $q = -0.4, \mu = 0.54, v = 1.8, f_0 = 0.046$ and $w = 1.1398$ with initial condition (-0.2309, 0.14602, 0) in Figure 7.3 and the corresponding time series plot for ϕ is presented in Figure 7.4 with fixed parametric values and initial condition, as in Figure 7.3. In this case, random sequences of uncorrelated motion of DIAWs are delineated. Thus, the plasma system produces quasiperiodic oscillation for the DIAWs.

Phase plot of the ADS (7.9) is delineated for $q = 1.4, \mu = 0.54, v = 1.8, f_0 = 0.9$ and $w = 1.9$ with initial condition (-0.2309, 0.14602, 0) in Figure 7.5 and the corresponding time series plot for ϕ is presented in Figure 7.6 with fixed parametric values and initial condition, as in

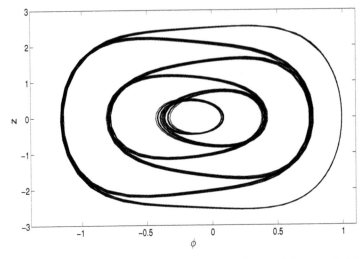

Figure 7.1: Phase plot of the ADS (7.9) for $q = 2$, $v = 0.8$, $\mu = 0.54$, $f_0 = 0.9$ and $w = 1.9$ with initial condition (-0.2309, 0.14602, 0).

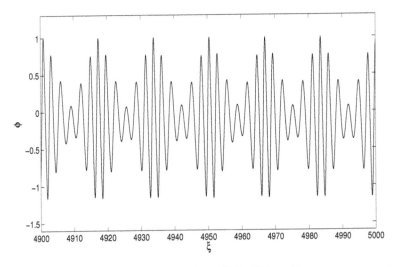

Figure 7.2: Time series plot of the ADS (7.9) with same parameters as Figure 7.1.

Figure 7.5. In this case, conservative chaotic oscillation of the DIAWs is discerned. The conservativeness of the chaotic motion for the DIAWs is confirmed by the nature of the Lyapunov exponents, presented in Figure 7.7.

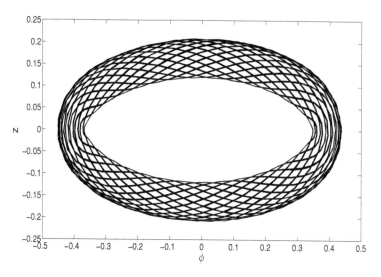

Figure 7.3: Phase plot of the ADS (7.9) for $q = -0.4, \mu = 0.54, v = 1.8$, $f_0 = 0.046$ and $w = 1.1398$ with initial condition (-0.2309, 0.14602, 0).

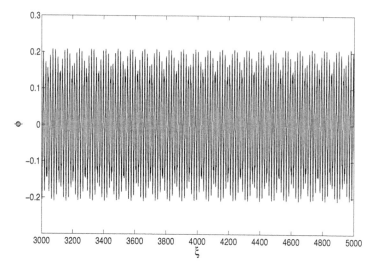

Figure 7.4: Time series plot of the ADS (7.9) with same parameters and initial condition as Figure 7.3.

Multiperiodic, quasiperiodic and chaotic oscillations of DIAWs are addressed successfully in an unmagnetized dusty plasma composed of cold inertial ions,

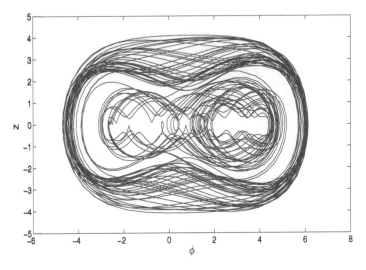

Figure 7.5: Phase plot of the ADS (7.9) for $q = 1.4, \mu = 0.54, \nu = 1.8, f_0 = 0.9$ and $w = 1.9$ with initial condition (-0.2309, 0.14602, 0).

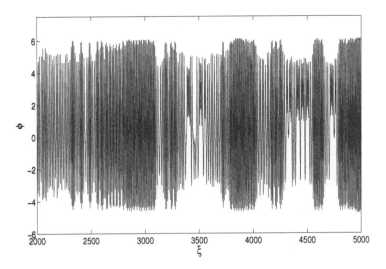

Figure 7.6: Time series plot of the ADS (7.9) with same parameters and initial condition as Figure 7.5.

negatively charged stationary dust particles and non-inertial q-nonextensive electrons.

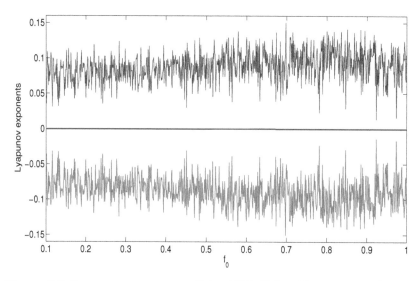

Figure 7.7: Lyapunov exponents of the ADS (7.9) with respect to f_0 for same values of other parameters and initial condition as Figure 7.5.

7.2 Multistability of electron-acoustic waves

Some nonlinear systems show the existence of more than one solution with particular set of parameters at different initial conditions. This interesting phenomenon is recognized as multistability and it significantly determines the nonlinear behaviors of physical systems [42]. Some experimental observations revealed that Q-switched gas laser supports the multistability property [43]. Multistability [44]-[49] is a new field of research in the nonlinear plasma systems and, thus, it requires intense exploration. The concept of planar dynamical systems is employed to examine multistability features of the electron-acoustic waves.

7.2.1 Basic equations

One-dimensional EAWs in a homogeneous plasma composed of mobile cold electrons, immobile ions, and q-nonextensive distributed hot electrons are considered. The basic equations for the dynamics of EAWs are:

$$\frac{\partial n}{\partial t} + \frac{\partial}{\partial x}(nu) = 0, \tag{7.10}$$

$$\frac{\partial u}{\partial t} + u\frac{\partial u}{\partial x} = \alpha\frac{\partial \phi}{\partial x}, \tag{7.11}$$

$$\frac{\partial^2 \phi}{\partial x^2} = n_e + \frac{1}{\alpha}n - (1 + \frac{1}{\alpha}), \tag{7.12}$$

where $\alpha = \frac{n_{e0}}{n_0} > 1$. Here, n (n_e) denotes number density of cold (hot) electrons. ϕ denotes electrostatic potential, u denotes velocity of cold electrons, T_h is temperature of hot electrons, m_h denotes electron mass, e is charge of electrons, and k_B is the Boltzmann constant. One can symbolize space variable and time by x and t, respectively. One can normalize n by n_0, u by $C_s = (k_BT_h/\alpha m_e)^{\frac{1}{2}}$, ϕ by k_BT_h/e, t by $\omega_{pc}^{-1} = (m_e/4\pi n_0 e^2)^{\frac{1}{2}}$ and x by $\lambda_{D-} = (k_BT_h/4\pi e^2 n_{e0})^{\frac{1}{2}}$.

The normalized hot electrons [41] number density is

$$n_e = [1 + (q-1)\phi]^{\frac{1}{q-1}+\frac{1}{2}}. \tag{7.13}$$

7.2.2 Multistability

The plasma system described by equations (7.10)-(7.13) is reduced to a coupled system (Refer Section 6.4 of Chapter 6) as

$$\begin{cases} \frac{d\Psi}{d\chi} = y, \\ \frac{dy}{d\chi} = M_1\Psi - M_2\Psi^3, \end{cases} \tag{7.14}$$

where $M_1 = \beta^2 - \frac{\beta v}{Pl^2}$, $M_2 = \frac{Q}{Pl^2}$, $P = -3C_1\frac{\omega^5}{2k^4}$, $Q = \frac{1}{12k^4\omega^5\beta_1}[18C_3k^4\omega^8\beta_1 + 4C_2\omega^8\beta_2 + 3k^8(2\beta^3 - \alpha^2\beta^4) + 12C_2k^4\beta^5\omega^4]$, with $\omega = \frac{k}{\sqrt{C_1+k^2}}$, $C_1 = \frac{q+1}{2}$, $C_2 = \frac{(q+1)(3-q)}{8}$, $C_3 = \frac{(1+q)(3-q)(5-3q)}{48}$, $\beta_1 = (3 - 3\omega^2 + \omega^4)$, $\beta_2 = (3 - 9\omega^2 + 5\omega^4)$, $\beta_3 = (9 - 9\omega^2 + 4\omega^4)$, $\beta_4 = (9 + 6\omega^2 - 12\omega^4 + 5\omega^6)$ and $\beta_5 = (-6 - \alpha\beta_1 + 4\omega^2)$.

The source perturbation can act as an external driving force in any nonlinear system. Recently, Mandi et al. [50] examined the nonlinear behavior of acoustic waves under the influence of periodic perturbation. Inclusion of an external periodic force of strength f_0 and frequency Ω in the form $f_0\cos(\Omega\chi)$ to the system (7.14) generates the following perturbed dynamical system

$$\begin{cases} \frac{d\Psi}{d\chi} = y, \\ \frac{dy}{d\chi} = M_1\Psi - M_2\Psi^3 + f_0\cos(\Omega\chi). \end{cases} \tag{7.15}$$

The nonautonomous system (7.15) can be expressed as the following autonomous system

$$\begin{cases} \frac{d\Psi}{d\chi} = y, \\ \frac{dy}{d\chi} = M_1\Psi - M_2\Psi^3 + f_0\cos(z), \\ \frac{dz}{d\chi} = \Omega. \end{cases} \tag{7.16}$$

The system (7.16) shows interesting types of coexisting trajectories which signify the presence of two or more types of motions of EAWs. In this work, variation of initial condition is taken under consideration to explore the existence of distinct trajectories in the system (7.16) with suitable fixed values of plasma parameters.

Existence of multistability behavior of EAWs for system (7.16) is presented in Figures 7.8-7.9 through phase space and time series plot. Coexistence of chaotic trajectory and quasiperiodic trajectory of electron-acoustic waves at initial conditions $(0.08, 0.01, 0)$ (blue) and $(0.089, 0.01, 0)$ (red), respectively, is shown in Figure 7.8 for $f_0 = 0.6$, $\Omega = 3$, $q = 0.55$, $l = 0.3$, $k = 0.3$, $\alpha = 1.5$, $\beta = 0.2$ and $v = 0.1$. Only the last 4000 iterations are taken under consideration out of total 20000 iterations for phase space presented in Figure 7.8. Time series plot for chaotic and quasiperiodic behaviours of electron acoustic wave is shown in Figure 7.9 with same initial and parametric conditions as Figure 7.8. In this case, iterations in between 19850 and 19900 are considered for presenting Figure 7.9.

To support the chaotic motion of the electron-acoustic waves, one can employ the most effective approach which can check the sensitivity of a system. If a system is bestowed with chaotic motion for any initial condition then one of its Lyapunov exponents shows a positive spectrum. For the system (7.16), external force's strength (f_0) and its frequency (Ω) modify the dynamics of EAWs significantly. Lyapunov exponents of the system (7.16) are presented with respect to f_0 in Figure 7.10 with the same initial and parametric conditions as in Figure 7.8. Figure 7.10 shows that the system (7.16) preserves its conservative property and the positive value of the Lyapunov exponent confirms the existence of chaotic motion for EAWs. Results of this work can be

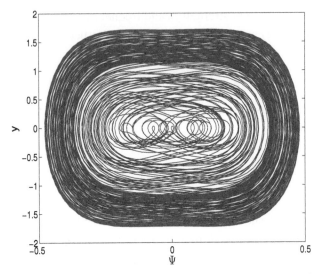

Figure 7.8: Chaotic trajectory at initial condition $(0.08, 0.01, 0)$ (blue) and quasiperiodic trajectory at initial condition $(0.089, 0.01, 0)$ (red) for $f_0 = 0.6$, $\Omega = 3$, $q = 0.55$, $l = 0.3$, $k = 0.3$, $\alpha = 1.5$, $\beta = 0.2$ and $v = 0.2$. In this case $M_1 = 0.483418$ and $M_2 = 12.5019$.

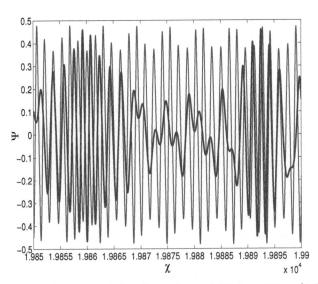

Figure 7.9: Time series plots of chaotic motion at initial condition $(0.08, 0.01, 0)$ (blue) and quasiperiodic motion at initial condition $(0.089, 0.01, 0)$ (red) for $f_0 = 0.6$, $\Omega = 3$, $q = 0.55$, $l = 0.3$, $k = 0.3$, $\alpha = 1.5$, $\beta = 0.2$ and $v = 0.2$. In this case $M_1 = 0.483418$ and $M_2 = 12.5019$.

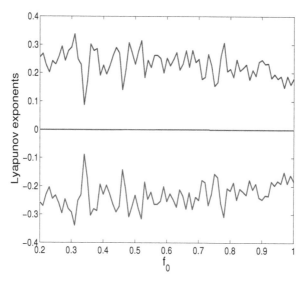

Figure 7.10: Plot of Lyapunov exponents with respect to f_0 for chaotic behavior with the same initial and parametric conditions as in Figure 7.8.

used to discern the nonlinear property of electron-acoustic wave structures in laboratory and astrophysical plasma environments.

7.3 Stable oscillation in a dissipative plasma

Propagation of nonlinear wave in dusty plasmas has gained a notable motivation by virtue of its wide applications in space plasmas [51], [52] and also in industrial plasmas [53]. Astrophysical and space plasmas contain ionized gases with neutral molecules, ions, electrons, and electrically charged massive dusts. The reports [54]-[56] show that the fluctuation of dust charge generates a damping of dust-acoustic waves (DAWs) [57] and dust-ion-acoustic waves in unmagnetized dusty plasmas. Recently, Gao [58] reported experimentally various acoustic waves in the cosmic or laboratory dusty plasmas. Using the concept of dynamical systems, stable oscillation of DAWs under the KdV-Burger equation in a nonextensive plasma with dust charge fluctuation is presented.

7.3.1 Model equations

A plasma system consisting of positive ions, electrons, negative ions and negatively charged dusts is considered. The density of electrons is comparatively higher than that of negative ions. Hence, q-nonextensive electrons along with Maxwellian negative and positive ions are considered in this system. The normalized basic equations for DAWs are given by

$$\frac{\partial n_d}{\partial t} + \frac{\partial n_d u_d}{\partial x} = 0, \tag{7.17}$$

$$\frac{\partial u_d}{\partial t} + u_d \frac{\partial u_d}{\partial x} = -\frac{Q}{\alpha_d} \frac{\partial \phi}{\partial x}, \tag{7.18}$$

$$\frac{\partial^2 \phi}{\partial x^2} = -\frac{\alpha_d}{(-\delta_n + \delta_p - 1)} [(-\delta_n + \delta_p - 1)Qn_d - \delta_n n_n - n_e + \delta_p n_p], \tag{7.19}$$

$$\frac{\omega_{pd}}{v_d} \left(\frac{\partial}{\partial t} + u_d \frac{\partial}{\partial x} \right) Q = \frac{1}{v_d e z_{d0}} (I_e + I_n + I_p), \tag{7.20}$$

where number density of q-nonextensive electrons is

$$n_e = (1 + (q-1)\phi)^{\frac{1}{q-1} + \frac{1}{2}}, \tag{7.21}$$

and number densities of Maxwellian negative and positive ions are given by

$$n_n = \exp\left(\frac{\phi}{\sigma_n}\right), \tag{7.22}$$

$$n_p = \exp\left(-\frac{\phi}{\sigma_p}\right). \tag{7.23}$$

The normalized current expressions are given by:

$$I_e = -\pi r_0^2 e n_{e0} \sqrt{\frac{8T_e}{\pi m_e}} B q [1 + (q-1)(ZQ + \phi)]^{\frac{1}{q-1} + 2}, \tag{7.24}$$

where

$$B_q = \begin{cases} \dfrac{(1-q)^{3/2}\Gamma(\frac{1}{1-q})}{q(2q-1)\Gamma(\frac{1}{q-1} - \frac{3}{2})}, & \text{for } -1 < q < 1, \\[3mm] \dfrac{(3q-1)(1-q)^{3/2}\Gamma(\frac{1}{q-1} + \frac{3}{2})}{2q(2q-1)\Gamma(\frac{1}{1-q})}, & \text{for } q > 1. \end{cases}$$

$$I_n = -\pi r_0^2 e n_{n0} \sqrt{\frac{8T_n}{\pi m_n}} n_n \exp\left(\frac{Z\phi}{\sigma_n}\right) \quad \text{and} \tag{7.25}$$

$$I_p = \pi r_0^2 e n_{p0} \sqrt{\frac{8T_p}{\pi m_p}} n_p \left(1 - \frac{ZQ}{\sigma_p}\right), \tag{7.26}$$

where n_d, n_n, n_p and n_e denote number densities of dusts, negative and positive ions, and electrons, respectively. The quasineutrality condition holds $n_{p0} = n_{n0} + n_{e0} + z_{d0}n_{d0}$, at the equilibrium condition. Here, u_d denotes dust velocity and ϕ denotes electrostatic wave potential. Here, $Z = e^2 z_{d0}/r_0 T_e$ and $Q = q_d/ez_{d0}$, where z_{d0} is unperturbed charge number settled on dust grains, q_d and r_0 denote charge and radius of grain.

The above physical variables are normalized as: n_d, n_n, n_p and n_e are normalized by n_{d0}, n_{n0}, n_{p0} and n_{e0}, respectively. u_d is normalized by $C_d = (z_{d0}T_e\alpha_d/m_d)^{\frac{1}{2}}$, where $\alpha_d = \frac{(1+\delta_n-\delta_p)}{(\delta_n/\sigma_n+(q+1)/2+\delta_p/\sigma_p)}$, m_d is dust mass. $\sigma_p = T_p/T_e$, $\sigma_n = T_n/T_e$, $\delta_p = n_{p0}/n_{e0}$, $\mu_n = n_{n0}/n_{e0}$, $\mu_p = n_{p0}/n_{e0}$. ϕ is normalized by T_e/e, time is normalized by $\omega_d^{-1} = (m_d/(4\pi n_{d0}z_{d0}^2 e^2))^{\frac{1}{2}}$, where ω_d is plasma frequency for dusts. x is normalized by the Debye length $\lambda_{D_d} = (T_e\alpha_d/(4\pi n_{d0}z_{d0}^2 e^2))^{\frac{1}{2}}$. The relation for the dust grain charging frequency is given by $v_d = -(\frac{\partial I_{tot}}{\partial q_d})_{eq}$. Now, δ_p is expressed as

$$\delta_p = \sqrt{\frac{\mu_p}{\sigma_p}} B_q \left[1 + C\frac{\exp(\frac{-Z}{\sigma_p})}{[1-(q-1)Z]^{\frac{2q-1}{q-1}}}\right]\left[\frac{[1-(q-1)Z]^{\frac{2q-1}{q-1}}}{(1+\frac{Z}{\sigma_p})}\right], \quad (7.27)$$

where $C = \delta_n\sqrt{\frac{\sigma_n}{\mu_n}}\frac{1}{B_q}$. Equation (7.27) is function of Z that is obtained from the equilibrium current balance condition $\frac{dq_d}{dt} = I_e + I_n + I_p = 0$, and considering $\phi = 0$ and $Q = -1$ in the expressions of I_e, I_n and I_p.

7.3.2   The KdV-Burgers equation

To explore small amplitude DAWs, the KdV-Burgers equation is derived by employing RPT. The space variable and time are stretched as

$$\xi = \varepsilon^{1/2}(x - \lambda t) \text{ and } \tau = \varepsilon^{3/2}t, \quad (7.28)$$

where ε denotes a small parameter that shows weakness of nonlinearity and λ is phase velocity of DAWs. Expansions of all dependent variables are expressed as

$$\begin{cases} n_d = 1 + \varepsilon n_{d1} + \varepsilon^2 n_{d2} + \cdots \\ u_d = 0 + \varepsilon u_{d1} + \varepsilon^2 u_{d2} + \cdots \\ \phi = 0 + \varepsilon\phi_1 + \varepsilon^2\phi_2 + \cdots \\ Q = -1 + \varepsilon Q_1 + \varepsilon^2 Q_2 + \cdots . \end{cases} \quad (7.29)$$

Substituting equations (7.21)-(7.29) in model equations (7.17)-(7.20) and comparing the coefficients of lowest order of ε, one can obtain

$$n_{d1} = \frac{u_{d1}}{\lambda}, \quad u_{d1} = -\frac{\phi_1}{\alpha_d \lambda}, \quad n_{d1} = -\frac{\phi_1}{\alpha_d \lambda^2} \quad \text{and} \quad Q_1 = \frac{1}{\alpha_d}\left(1 - \frac{1}{\lambda^2}\right)\phi_1. \quad (7.30)$$

Comparing the coefficients of next higher-order of ε, one can obtain

$$\frac{\partial n_{d1}}{\partial \tau} - \lambda\frac{\partial n_{d2}}{\partial \xi} + \frac{\partial u_{d1}}{\partial \xi} + \frac{\partial n_{d1}u_{d1}}{\partial \xi} = 0, \quad (7.31)$$

$$\frac{\partial u_{d1}}{\partial \tau} - \lambda\frac{\partial u_{d2}}{\partial \xi} + u_{d1}\frac{\partial u_{d1}}{\partial \xi} = -\frac{1}{\alpha_d}\left(Q_1\frac{\partial \phi_1}{\partial \xi} - \frac{\partial \phi_2}{\partial \xi}\right), \quad (7.32)$$

$$\frac{\partial^2 \phi_1}{\partial \xi^2} = \alpha_d n_{d2} - \alpha_d Q_2 + \phi_2 + E\phi_1^2, \quad (7.33)$$

with $\lambda = \sqrt{\dfrac{1}{1+\alpha_d\beta_d}}$ and $E = \dfrac{1}{\lambda^2\alpha_d}\left(1 - \dfrac{1}{\lambda^2}\right) - \dfrac{1}{\lambda^2}\dfrac{1}{2\left(\frac{q+1}{2} + \frac{\delta_n}{\sigma_n} + \frac{\delta_p}{\sigma_p}\right)}\left(\frac{\delta_p}{\sigma_p^2} - \right.$

$\left.\frac{1}{4}(q+1)(3-q) - \frac{\delta_n}{\sigma_n^2}\right)$. Moreover, in the case of adiabatic dust charge variation, the dust charging frequency is much greater than the dusty plasma frequency, this results in $\frac{\omega_{pd}}{\nu_d} \approx 0$, thus, the grain charging equation becomes $I_e + I_n + I_p = 0$. From this relation, it is easy to obtain the following expressions for Q_1 and Q_2 as

$$Q_1 = -\beta_d\phi_1, \quad Q_2 = -\beta_d\phi_2 + r_d\phi_1^2, \quad (7.34)$$

where $\beta_d = \dfrac{\left[\frac{A}{\sigma_p}\left(1 + \frac{Z}{\sigma_p}\right) + B(2q-1) + \frac{B_1C}{\sigma_n}\right]}{[(2q-1)(1-qZ)Z + \frac{AZ}{\sigma_p} + \frac{CZ}{\sigma_n}(1 - \frac{Z}{\sigma_n})]}$,

$r_d = -\dfrac{\left[-\frac{A}{2\sigma_p^2}\left(1 + \frac{Z}{\sigma_p}\right) + \frac{A\beta_d Z}{\sigma_p^2} + \frac{1}{2}q(2q-1)(1-\beta_d Z)^2(1-Z) + \frac{C\beta_d^2 Z^2}{2\sigma_n^2} + \frac{B_1C}{2\sigma_n^2} - (1 - \frac{Z}{\sigma_n})\frac{C\beta_d Z}{\sigma_n^2}\right]}{[(2q-1)(1-qZ)Z + \frac{CZ}{\sigma_n}(1 - \frac{Z}{\sigma_n}) + \frac{AZ}{\sigma_p}]}$,

$A = \delta_p\sqrt{\dfrac{\sigma_p}{\mu_p}}\dfrac{1}{B_q}$, $B = 1 - \dfrac{Z}{\sigma_n} + \dfrac{Z^2}{2!\sigma_n^2}$ and $B_1 = 1 - qZ + \dfrac{1}{2}qZ^2$.

Removing all higher order perturbed terms from equations (7.31)-(7.34), the KdV equation is obtained as,

$$\frac{\partial \phi_1}{\partial \tau} + a\phi_1\frac{\partial \phi_1}{\partial \xi} + b\frac{\partial^3 \phi_1}{\partial \xi^3} = 0, \quad (7.35)$$

where $a = \alpha_d b\left[2r_d - \dfrac{3}{\alpha_d^2\lambda^4} + \dfrac{\beta_d}{\alpha_d\lambda^2} - \dfrac{2E}{\alpha_d}\right]$ and $b = \dfrac{1}{2(1+\alpha_d\beta_d)^{3/2}}$.

For non-adiabatic case of dust charge variation, i.e., for slow charging process, we consider $\frac{\omega_{pd}}{v_d} = v\sqrt{\varepsilon}$, where v is small value of order unity and represents the strength of nonadiabaticity. Then, the coefficients of ε and ε^2 give

$$Q_1 = -\beta_d\phi_1, \quad Q_2 = -\beta_d\phi_2 + r_d\phi_1^2 + \mu_d\frac{\partial\phi_1}{\partial\xi}. \tag{7.36}$$

Eliminating all higher-order perturbed terms from equations (7.31)-(7.36), the KdV-Burger equation is obtained as

$$\frac{\partial\phi_1}{\partial\tau} + a\phi_1\frac{\partial\phi_1}{\partial\xi} + b\frac{\partial^3\phi_1}{\partial\xi^3} = \mu\frac{\partial^2\phi_1}{\partial\xi^2}, \tag{7.37}$$

where $\mu = -\frac{\mu_d\lambda^3\alpha_d}{2}$ is the coefficient of dissipative viscous effect generated by the dust charge variation by virtue of non-adiabaticity, while $\mu_d = v\frac{\mu_{d1}}{\mu_{d2}}\mu_{d3}$, with

$$\mu_{d1} = \lambda\beta_d(2q-1)Z[1-(q-1)Z]^{\frac{1}{q-1}+1}, \qquad \mu_{d2} = \left[(2q-1)(1-qZ)Z + \right.$$

$$\frac{CZ}{\sigma_n}\left(1-\frac{Z}{\sigma_n}\right) + \frac{AZ}{\sigma_p}\right] \text{ and}$$

$$\mu_{d3} = \left[1+\frac{C}{\sigma_n(2q-1)}\exp\left(-\frac{Z}{\sigma_n}\right)[1-(q-1)Z]^{-\left(\frac{1}{q-1}+1\right)} + \frac{A}{\sigma_p(2q-1)}[1-\right.$$

$$(q-1)Z]^{-\left(\frac{1}{q-1}+1\right)}\right].$$

7.3.3 Stability analysis of DAWs

The characteristic behaviors of DAWs are investigated with the consideration of a very fundamental concept of dynamical system (DS) through phase profiles. The purpose of presenting phase profiles is to show the stability of the dynamical system obtained from the model equations. The phase plane analysis in the theory of nonlinear DS is one of the most appropriate approaches to study behaviors of DS. In particular, qualitative phase portraits and time series plots are obtained by varying the physical parameters. Therefore, to investigate the stability analysis of the KdV-Burgers equation (7.37), one can employ qualitative techniques of the planar DS, such as, phase portrait analysis and time series plots. One can transform equation (7.37) using $\eta = \xi - V\tau$

into the following nonlinear planar DS

$$-V\frac{d\phi_1}{d\eta} + a\phi_1\frac{d\phi_1}{d\eta} + b\frac{d^3\phi_1}{d\eta^3} - \mu\frac{d^2\phi_1}{d\eta^2} = 0, \tag{7.38}$$

where V is velocity of wave. Integrating equation (7.38) with respect to η and applying the boundary conditions as $|\eta| \to \infty$, $n_d \to 1, u_d$ and $\phi_1 \to 0$, it is easy to obtain

$$\frac{d^2\phi_1}{d\eta^2} = \frac{V}{b}\phi_1 - \frac{a}{2b}\phi_1^2 + \frac{\mu}{b}\frac{d\phi_1}{d\eta}. \tag{7.39}$$

Equation (7.39) is represented as the following DS:

$$\begin{cases} \dfrac{d\phi_1}{d\eta} = z, \\ \dfrac{dz}{d\eta} = \dfrac{V}{b}\phi_1 - \dfrac{a}{2b}\phi_1^2 + \dfrac{\mu}{b}z. \end{cases} \tag{7.40}$$

Equation (7.40) has two fixed points at $(\phi_1, z) = (0,0)$ and $(\phi_1, z) = (2V/a, 0)$, where the fixed point $(0,0)$ is a saddle point and there is a stable spiral at $(2V/a, 0)$. The qualitative changes of the system (7.40) are examined for system parameters q and v. Since the coefficient of dissipative viscosity (μ) is generated by the non-adiabaticity (v) of the dust charge variation, it is required to examine the dynamics of the DAWs by varying v.

In Figures 7.11 and 7.12, phase plots of the DS (7.40) are discerned for $v = 0.1$ and $v = 0.15$, respectively, with fixed values of other parameters $\delta_n = 0.02$, $q = 1.2$, $\sigma_n = 0.3$, $\sigma_p = 0.03$, $\mu_n = 0.3$, $\mu_p = 0.4$, and $V = 0.91$. In Figures 7.13 and 7.14, corresponding time series plots of the stable oscillation of dust-acoustic wave are delineated for the phase plots presented in Figures 7.11 and 7.12. It is observed from Figures 7.13 and 7.14 that the oscillations of DAWs decay rapidly for increasing in v.

Thus DAW features are investigated in a four-component dusty plasma composing of Maxwellian distributed positive and negative ions, and q-nonextensive distributed electrons. Here, DAWs are investigated in the presence of dust charge variation. The KdV-Burgers (KdV-B) equation is derived by applying multi-scale RPT. Using the traveling wave frame, the KdV-Burgers is transformed to a DS. With the help of phase plane analysis, nonlinear stable spiral structures of the DAWs are reported by varying strength of non-adiabaticity parameter v. It is important to note that the oscillation

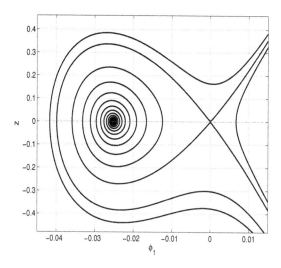

Figure 7.11: Phase plot of the DS (7.40) for $v = 0.1$, $\delta_n = 0.02$, $q = 1.2$, $\sigma_n = 0.3$, $\sigma_p = 0.03$, $\mu_n = 0.3$, $\mu_p = 0.4$, and $V = 0.91$.

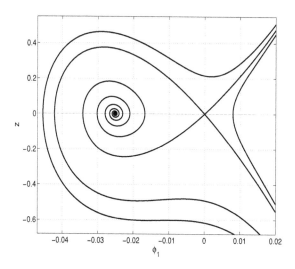

Figure 7.12: Phase plot of the DS (7.40) for $v = 0.15$, $\delta_n = 0.02$, $q = 1.2$, $\sigma_n = 0.3$, $\sigma_p = 0.03$, $\mu_n = 0.3$, $\mu_p = 0.4$, and $V = 0.91$.

of DAWs decays rapidly for increasing v with fixed values of other physical parameters.

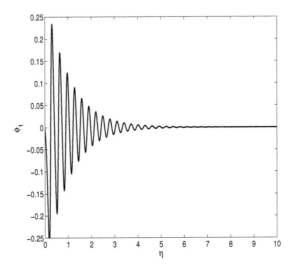

Figure 7.13: Stable oscillation of dust-acoustic wave for $v = 0.1$, $\delta_n = 0.02$, $q = 1.2, \sigma_n = 0.3$, $\sigma_p = 0.03$, $\mu_n = 0.3, \mu_p = 0.4$, and $V = 0.91$.

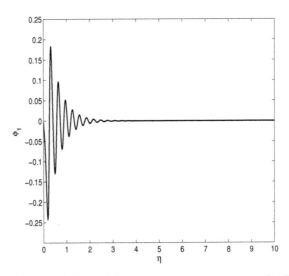

Figure 7.14: Stable oscillation of dust-acoustic wave for $v = 0.15$, $\delta_n = 0.02$, $q = 1.2, \sigma_n = 0.3$, $\sigma_p = 0.03$, $\mu_n = 0.3, \mu_p = 0.4$, and $V = 0.91$.

Dynamical features, such as chaos of arbitrary amplitude DIAWs and multi-stability of finite amplitude EAWs, are investigated in different plasmas in the presence of an external periodic force using efficient tools like phase profiles, time series graphs and Lyapunov exponent spectrum. Another dynamical feature, namely stable oscillation of DAWs is also investigated in the framework of KdV-Burgers equation in a nonextensive plasma with dust-charge variation. While chaos and multistability are significantly affected by the strength and frequency of the external force, stable oscillations are notably influenced by nonextensive parameter and strength of nonadiabaticity parameter. All these features occur as an effect of nonlinearity and are quite essential in the study of dynamical systems and nonlinear waves in plasmas.

References

[1] P. K. Shukla and A. A. Mamun, Introduction to Dusty Plasma Physics, Institute of Physics, Bristol (2002).

[2] F. Verheest, Waves in Dusty Plasmas, Kluwer Academic, Dordrecht (2000).

[3] P. K. Shukla and V. P. Silin, Phys. Scr., 45: 508 (1992).

[4] N. N. Rao, P. K. Shukla and M. Y. Yu, Planet Space Sci., 38: 543 (1990).

[5] I. Kourakis and P. K. Shukla, Eur. Phys. J. D, 30: 97 (2004).

[6] F. Melandso, Phys. Plasmas, 3: 3890 (1996).

[7] P. K. Shukla and R. K. Varma, Phys. Fluids B, 5: 236 (1993).

[8] M. Tribeche and T. H. Zerguini, Phys. Plasmas, 11: 4115 (2004).

[9] A. A. Mamun and P. K. Shukla, Phys. Plasmas, 9: 1468 (2002).

[10] H. Alinejad, Astrophys. Space Sci., 327: 131 (2010).

[11] W. F. Ei-Taibany, N. A. Ei-Bedwely and E. F. Ei-Shamy, Phys. Plasmas, 18: 033703 (2011).

[12] A. Barkan, N. D'Angelo and R. L. Merlino, Planet. Space Sci., 44: 239 (1996).

[13] U. K. Samanta, A. Saha and P. Chatterjee, Physics of Plasmas, 20: 022111 (2013a).

[14] U. K. Samanta, A. Saha and P. Chatterjee, Astrophysics and Space Science, 347: 293 (2013b).

[15] A. Saha and P. Chatterjee, Astrophysics and Space Science, 349: 813–820 (2014).

[16] T. K. Das, A. Saha, N. Pal and P. Chatterjee, Physics of Plasmas, 24: 073707 (2017).

[17] Y. Nakamura, H. Bailung and P. K. Shukla, Phys. Rev. Lett., 83: 1602 (1999).

[18] Y. Nakamura and A. Sarma, Phys. Plasmas, 8: 3921 (2001).

[19] M. G. M. Anowar and A. A. Mamun, Physics Letters A, 372: 5896 (2008).

[20] P. K. Shukla, N. N. Rao, M. Y. Yu and N. L. Tsintsa, Phys. Rep., 135: 1 (1986).

[21] H. R. Pakzad, Phys. Lett. A, 373: 847 (2009).

[22] M. Sakagami and A. Taruya, Contin. Mech. Thermodyn., 16: 279 (2004).

[23] F. Nobre and C. Tsallis, Physica A, 213: 337 (1995).

[24] C. Tsallis, J. Stat. Phys., 52: 479 (1988).

[25] A. Renyi, Acta Math. Hung., 6: 285 (1955).

[26] M. Bacha and M. Tribeche, Astrophys. Space Sci., 337: 253 (2012).

[27] H. R. Pakzad and M. Tribeche, Astrophys. Space Sci., 334: 45 (2011).

[28] B. Sahu and M. Tribeche, Astrophys. Space Sci., 338: 259 (2012).

[29] K. Nozaki and N. Bekki, Phys. Rev. Lett., 50: 1226 (1983).

[30] G. P. Williams, Chaos Theory Tamed, Washington, Joseph Henry (1997).

[31] W. Beiglbock, J. P. Eckmann, H. Grosse, M. Loss, S. Smirnov, L. Takhtajan and J. Yngvason, Concepts and Results in Chaotic Dynamics, Springer, Berlin (2000).

[32] B. Sahu, S. Poria, U. N. Ghosh and R. Roychoudhury, Phys. Plasma, 19: 050326 (2012).

[33] B. Sahu, S. Poria and R. Roychoudhury, Astrophys. Space Sci., 341: 567 (2012).

[34] H. Zhen, B. Tian, Y. Wang, H. Zhong and W. Sun, Phys. Plasma, 21: 012304 (2014).

[35] H. Zhen, B. Tian, Y. Wang, W. Sun and L. Liu, Phys. Plasma, 21: 073709 (2014).

[36] A. Roy, A. P. Misra and S. Banerjee, Optik, 176: 119–131 (2019).

[37] P. K. Prasad, A. Gowrisankar, A. Saha and S. Banerjee, Physica Scripta, 95(6): 065603 (2020).

[38] A. Roy, A. P. Misra and S. Banerjee, Physica Scripta, 95(4): 045225 (2020).

[39] A. Saha, B. Pradhan and S. Banerjee, The European Physical Journal Plus, 135(2): 216 (2020).

[40] A. Saha and P. Chatterjee, The European Physical Journal D, 69: 203 (2015).

[41] A. S. Bains, M. Tribeche and T. S. Gill, Phys. Plasmas, 18: 022108 (2011).

[42] H. Natiq, S. Banerjee, A. Misra and M. Said, Chaos, Solitons & Fractals, 122: 58 (2019).

[43] F. Arecchi, R. Meucci, G. Puccioni and J. Tredicce, Physical Review Letters, 49: 1217 (1982).

[44] A. Saha, J. Tamang, G. C. Wu and S. Banerjee, Communications in Theoretical Physics, 72(11): 115501 (2020).

[45] H. Natiq, M. R. M. Said, M. R. K. Ariffin, S. He, L. Rondoni and S. Banerjee, The European Physical Journal Plus, 133(12): 557 (2018).

[46] B. Yan, P. K. Prasad, S. Mukherjee, A. Saha and S. Banerjee, Complexity, 2020 (2020).

[47] A. Saha, S. Sarkar, S. Banerjee and K. K. Mondal, The European Physical Journal Special Topics, 229: 979–988 (2020).

[48] A. Saha, B. Pradhan and S. Banerjee, Physica Scripta, 95(5): 055602 (2020).

[49] B. Pradhan, S. Mukherjee, A. Saha, H. Natiq and S. Banerjee, Zeitschrift für Naturforschung A, 76(2): 109–119 (2021).

[50] L. Mandi, A. Saha and P. Chatterjee, Advances in Space Research, 64: 427 (2019).

[51] M. Horanyi and D. A. Mendis, Astrophys J., 294: 357 (1985).

[52] C. K. Goertz, Rev. Geophys., 27: 271 (1989).

[53] V. E. Fortov, A. V. Ivlev, S. A. Khrapak, A. G. Khrapak and G. E. Morfill, Phys. Rep, 421: 1 (2005).

[54] R. K. Varma, P. K. Shukla and V. Krishan, Phys. Rev. E, 47: 3612 (1993).

[55] C. Cui and J. Goree, IEEE Transactions on Plasma Science, 22: 151 (1994).

[56] F. Melandso, T. K. Aslaksen and O. Havnes, Planet. Space Sci., 41: 321–325 (1993).

[57] N. N. Rao, P. K. Shukla and M. Y. Yu, Planet Space Sci., 38: 543–546 (1990).

[58] X. Gao, Applied Mathematics Letters, 91: 165 (2018).

Index

Printed and bound by CPI Group (UK) Ltd, Croydon, CR0 4YY

17/10/2024

01775709-0002